Viscosity and Diffusivity

VISCOSITY AND DIFFUSIVITY
A Predictive Treatment

JOEL H. HILDEBRAND

Department of Chemistry
University of California, Berkeley

A WILEY-INTERSCIENCE PUBLICATION

JOHN WILEY & SONS, New York ● London ● Sydney ● Toronto

Library of Congress Cataloging in Publication Data

Hildebrand, Joel Henry, 1881-
    Viscosity and diffusivity.

    "A Wiley-Interscience publication."
    1. Viscosity. 2. Diffusion. I. Title.

QD543.H52   532'.0533   77-2383
ISBN 0-471-03072-4

Printed in the United States of America

10 9 8 7 6 5 4 3 2 1

Foreword

There is much to be learned from this compact little monograph. Joel Hildebrand is a truly great scientist and a student of nature. For the past 60 years, since his famous paper on "Solubility," Hildebrand has been trying to understand the properties of liquids. Somehow, he has the ability to sweep away all the complexities and discover simple relationships that will take theoreticians another generation to derive. The secret of Hildebrand's success is that he is a very ingenious and tireless experimenter who demands perfection and intellectual honesty of himself and his colleagues. His mind is not clouded by scientific dogma and, when the experiments dictate, he makes statements which Gilbert Lewis would call "impertinent but also pertinent."* To Hildebrand, *CHEMISTRY IS FUN!* Throughout his writings, one is impressed by his enthusiasm and by his "reverence for the profundity, subtlety, and beauty of nature."*

The readers of this monograph cannot fail to be impressed by the slick empirical relations that Hildebrand has discovered—far simpler and more accurate than other formulae which are widely used. He explained to me that "every formula must express a clear concept of the physics of a process, and must be tested by seeking nature's opinion through carefully designed experiments or appropriate plots of experimental data. A formula or model must be at least qualitatively consistent with *all* pertinent properties. If the formula definitely conflicts with any one of the facts, the formula impedes progress and should be discarded." In other words, Hildebrand agrees with Wigner's statement that "a wastebasket is a scientist's best friend!"

---

*Hildebrand, *Chem. Eng. News,* Sept. 13, 1976, p. 25.

The publication of this monograph coincides with Hildebrand's ninety-fifth birthday. Congratulations, Joel, for a long and useful career. You have taught three generations of chemists the true meaning of science. You have taught us to respect "the power and beauty, not just the utility, of unrestricted human thought." We greatly admire you as a great scientist, as a wonderful friend, and most of all, we admire the way in which you have strived to make this a better world in which to live. It is truly amazing that at the age of 95, your mind is still crystal clear and your research becomes progressively more fascinating. We hope that you will continue to persevere for many healthy, happy years to come!

JOSEPH O. HIRSCHFELDER

Madison, Wisconsin
December 6, 1976

Preface

This monograph consists of thirteen papers on viscosity, diffusivity, and solubility written during the years 1971 to 1976. Twelve of them were published in several different journals because they deal with properties that interest different groups of readers: biologists, engineers, theorists on the liquid state, and authors of textbooks. Five of them appeared in the *Proceedings of the National Academy of Sciences,* where publication takes only three months and where it is not necessary to argue with a referee. Some years ago an editor wrote that he hoped I would be able to revise a certain paper to meet the objection of a referee who had reported: "Something must be terribly wrong; Hildebrand presents a view that is contrary to what everybody else thinks."

I submitted paper number 12 of this book, as an experiment, to a certain journal of physics. It was returned to me with a recommendation that it be submitted to a journal of chemical physics because it is "empirical," and hence not suitable for their journal. My dictionary defines that word as "based upon experiment or observation," which, if ruled out, would seem to leave only metaphysics.

The original papers were well received by the scientific and technical public; hundreds of reprints were requested, and I decided that they should be available together in a monograph, since they are interdependent. They present for the first time a predictive treatment of "transport properties."

A freshman, once upon a time, asked me to define chemistry; I replied, "Chemistry is what chemists do, and how they do it." I believe a reader of the book will get more out of it if I tell a little about what I did, and how I did it.

Ever since my early childhood I have been curious about natural phenomena. By the time I was 12 years old I had a workshop and laboratory. In high school I demonstrated that the formula of nitric oxide is NO, not N_2O_2, as asserted in a book by a professor of chemistry at Harvard. In a book by P. Rothmud written in 1907, titled *Löslichkeit,* he asserted in his Foreword that "there is a long way to go before we will have built a unified theory of solubility comparable to our knowledge of electrochemistry."

That was a challenge, so in 1916 I published the germ of a general theory. The problem was complex, and years passed before a theory of regular solutions attained a degree of maturity. The steps were outlined in 1971 in a paper in *Science* titled "Order from Chaos." The norm is a mixture of two liquids composed of nonpolar, approximately symmetrical molecules in liquids with no "structure" to resist even an infinitessimal stress. All the molecules of such a liquid are in a continuous state of thermal agitation, with maximum entropy, or disorder. This concept is basic for the theory of regular solutions, and it works, as can be seen in a plot of the two members of the regular solution equation using solutions of iodine in 35 liquids, as shown on p. 148 in *Regular and Related Solutions,* by Hildebrand, Prausnitz, and Scott, Van Nostrand Reinhold Co., 1970. One member of the equation is a function of the solubility of iodine; the other is the theoretical expression for the attractive potential between iodine and solvent. Points for 14 solvents in which color is violet, like that of the vapor of iodine, are on a 45° line, indicating agreement of theory with fact. Points for polar liquids deviate from this line. Aromatic solvents, which form complexes with iodine, with colors from red to brown, deviate from the line increasingly from benzine to mesitylene, as expected. Alkanes give violet solutions, but the points deviate from the "regular" line in proportion to the ratios of methyl to methylene. There is as yet no theory of attractive potential between molecules with outer, nonbonded electrons and molecules with electrons only in bonds.

The fact that the theory of regular solutions can deal so comprehensively with the solubility of iodine in 35 liquids is evidence that the concepts of the liquid state on which it was based are probably valid. I was puzzled by the fact that published papers and books contained models of liquids with "liquid-lattices," "clusters," "monomers and dimers," "cells," "cages," and "significant structures," all picked out of the sky to furnish adjustable parameters. I consulted nature by means of experiments and data plotted in different ways.

In 1915 I had plotted vapor pressures of nonpolar liquids as $\log P$ against $\log T$ instead of as $\log P$ against $1/T$, the "regular" way.

The former method gives the entropy of vaporization, the latter the heat of vaporization. The difference in entropy between liquids and gases all at the same volumes measures the entropy of the liquid. Hermsen and Prausnitz determined the entropy of vaporization of 20 alkanes whose boiling points range from 92.6K to 611.9K, all to gases at 49.5 liters. It was 22.3 ± 0.1 cal deg^{-1} mole. $^{-1}$ Any order in a liquid imposed by dipoles or hydrogen bonds that must be broken in vaporization adds to the entropy of vaporization by different amounts; therefore, minimum entropy indicates zero order, maximum disorder, and thus definitely rules out anything like "liquid lattices."

Progress in understanding nature has been achieved largely by persons who are not bound by "what everybody else thinks." Let me illustrate. In 1956 I could see no reason why, if there were a hole of molecular size, an adjacent molecule could not freely move into it. My frequent consultant, Dr. Berni J. Alder, and I decided to ask nature whether we might be wrong, so my experimentalist, Dick Haycock, determined the diffusivity of iodine in carbon tetrachloride over a range of temperature, first at 1 atm, then at 200 atm, where, presumably, "holes" would be fewer.

That was in 1953, while I was still under the spell of orthodoxy, according to which the data should be plotted as log D against $1/T$, with the slope of the line giving an "activation energy" which implied a "potential barrier." But now, in 1976, I have become a heretic; I simply plot D against T. I thus obtain two parallel lines, straighter than those of 1953. Diffusivity increases linearly with temperature at constant pressure because the liquid expands and the ratio of unoccupied to occupied volume increases. But at 200 atm it is smaller. It is that simple—and there are no holes in the argument.

The difference between the way I treated the data in 1954 and in 1976 illustrates the differences between the transport theory of the books and the theory that is presented in this monograph. A few years ago a mathematician was included in a symposium on transport processes to tell the participants how to get physics out of mathematics. He didn't. Some theorists appear to have been trying to learn about liquids by the process advised by Plato for learning astronomy:

And let us dismiss the heavenly bodies, if we intend truly to apprehend astronomy, and render profitable instead of unprofitable that part of the soul which is naturally wise.

Alfred North Whitehead wrote:

There can be no true physical science which looks to mathematics for the
provision of a conceptual model. Such a procedure is to repeat the errors of
the logicians of the Middle Ages.

The papers collected into this small monograph were all written
from the standpoint of simple physics; the quantitative formula-
tions that emerged are of the simplest sort, but they yield agreement with experiment to three significant figures.

In the first paper I tell what I learned by abandoning exponen-
tials, the orthodox way, and simply plotting diffusivity against
temperature, and fluidity (the reciprocal of viscosity) against liquid
molal volume, as it varies with temperature. I learned that diffusion should not be treated as Stokes law sedimentation, caused by
a vector force acting against a coefficient of friction, but that it
results simply from thermal motions that prevent any molecule
from staying "put." Self-diffusion can occur in liquids expanded so
little from their minimum volume that mean free paths are only
small fractions of molecular diameters.

I found that fluidity is a linear function of molal volume except
at very small volumes, where the freedom of rotation of unsymmetrical molecules may be inhibited. I found also that the linear
relation can hold to high pressures. However, as volumes approach
critical volumes, and molecules, instead of remaining in the fields
of force of many near neighbors, begin to have binary collisions and
paths long enough for them to acquire significant fractions of their
mean momentum in free flight, the straight lines split into
isotherms and curve toward zero slope. Vector momentum applied
to produce flow is opposed by the randomly oriented thermal
momenta that account for the viscosity of dilute gases. I discovered
the form of the fraction of the free space momentum that permits
accurate calculation of the viscosity of compressed fluids, a "first"
in kinetic theory.

The formula for calculating the viscosity of dilute gases was
derived for hard spheres. I found how to modify it to yield correct
viscosities for gases and vapors from neon to chloroform.

I wondered whether liquid metals would conform to my procedure and had the satisfaction of finding that they do. Their intrinsic volumes per mole from crystal densities at ordinary temperatures are virtually identical to those obtained from fluidities of
liquids at high temperatures, and their large range of viscosities at

equal degrees of expansion are consistent with electron structures.

The word *entropy* occurs several times in the foregoing. Many scientists seem to have little more than a speaking acquaintance with the term. An instructor in engineering, when asked by a student what entropy is, replied, "I don't know what entropy is, nobody knows what it is; you just use it." I offer an additional illustration of the value of acquaintance with its statistical nature.

Paper number 7 in this book does not mention either viscosity or diffusivity, but it shows that when molecules with small attractive potential are dissolved in a liquid such as carbon tetrachloride, there is considerable expansion and, therefore, increase in entropy. The increased freedom of motion of the solute molecule and its immediate neighbors contributes notably to its diffusivity, as shown in paper number 8. This could hardly have been discovered except from solubility over a range of temperature.

I have greatly enjoyed writing these papers. It has been exciting to have questions answered, again and again, by straight lines showing agreement between fact and theory. The experience has been like that described by Henri Poincaré when he wrote:

The scientist does not study nature because it is useful; he studies it because he delights in it, and he delights in it because it is beautiful. Of course, I do not here speak of that beauty which strikes the senses, the beauty of qualities and of appearances; not that I undervalue such beauty, far from it, but it has nothing to do with science; I mean that profounder beauty which comes from the harmonious order of the parts and which a pure intelligence can grasp. This it is which gives body, a structure so to speak, to the iridescent appearances which flatter our senses, and without this support the beauty of these fugitive dreams would be only imperfect, because it would be vague and always fleeting. On the contrary, intellectual beauty is sufficient unto itself, and it is for its sake, more perhaps than for the future good of humanity, that the scientist devotes himself to long and difficult labors.

I wish here to state my gratitude to the collaborators who applied their skill and intelligence to obtaining valid answers to the questions we have formulated, especially to Robert H. Lamoreaux, coauthor of six of these papers; also to my continual consultant, Berni J. Alder. The process has been another case of getting "order from chaos." My satisfaction is crowned by the generous appraisal of the work by my longtime friend, Joe Hirschfelder in his Foreword. He was senior editor of the monumental book, *Kinetic Theory of Gases and Liquids,* published

likewise by John Wiley & Sons, New York, in 1946. It has long
been the authoritative work in this field. His opinion carries more
weight with me than all that reviewers may write about this
monograph.

JOEL H. HILDEBRAND

Berkeley, California
February 1977

Contents

Viscosity and Diffusivity

1

Motions of Molecules in Liquids: Viscosity and Diffusivity

Joel H. Hildebrand

Abstract. *The fluidity of a simple liquid is proportional to its degree of expansion over the volume, V_0, at which its molecules are so crowded as to inhibit self-diffusion and viscous (as distinguished from plastic) flow. The equation of proportionality is $1/\eta = B[(V - V_0)/V_0]$ where η is the viscosity and V is the molal volume. Values of B are the same for normal paraffins from C_3H_8 to C_7H_{16} and then decrease progressively as the paraffin lengths increase. Values for other liquids, C_6H_6, CCl_4, P_4, CS_2, $CHCl_3$, and Hg, appear to vary with repulsive forces. Liquids can be moderately fluid when expanded by less than 10 percent; this shows the unreality of some theoretical treatments of the liquid state. Diffusivity begins from the temperature at which V equals V_0 and can be correlated for temperature dependence, and for solute-solvent interrelations.*

Bird, Stewart, and Lightfoot had this to say in chapter 16 of their extraordinarily fine book on *Transport Phenomena* about the present state of theory of diffusion in liquids (*1*):

If the reader has by now concluded that little is known about the prediction of dense gas and liquid diffusivities, he is correct. There is an urgent need for experimental measurements, both for their own value and for the development of future theories.

I see two reasons for this lack of general theory. One, measurements have not been designed to obtain answers to questions of general significance. Although, for example, diffusion is an example of entropy increasing toward the maximum permissible under the conditions, very few of the vast number of measurements of diffusivity (*2*) were made at more than one temperature. Also, statistical mechanicians have treated transport theory as essentially a problem in mathematics. In papers and symposia dealing with the

Reprinted from

29 October 1971, volume 174 pages 490-493

theory of the liquid state, few of the authors have sought validation by experimental facts. The volume recording the proceedings of the International Symposium on Transport Processes in Statistical Mechanics (3), held in Brussels in 1956, is typical. In the first half of the book there is only one paper containing any reference to experiment. Otto Redlich has said that science, unlike mathematics, is not autonomous; its concepts must be referred to nature for validation. Theorists do not always do that.

I approach the problem of diffusion by way of viscosity, the much simpler phenomenon. Viscosity of liquids has been treated copiously in engineering and scientific literature, but nearly all that I have read seems unrealistic in one respect or another, such as the assumption of quasi-lattice structure that ignores clear evidence to the contrary, or that temperature dependence is exponential, or that there is an energy of activation, a notion that disregards the basic distinction between liquid and plastic flow.

Batschinski (4), in 1913, published an important paper that has been virtually ignored by authors on transport phenomena; I found it only recently, almost by accident. Batschinski reasoned that viscosity is not a direct function of temperature but of the difference between the specific volume of the liquid, v, and a certain constant, ω, similar to the van der Waals b. He wrote his relation $\eta = c/(v - \omega)$, where c is another constant. He plotted v against fluidity, $1/\eta$, for 87 liquids, obtaining straight lines for those that are not associated. The values he thus obtained for ω fell between specific volumes of liquid and solid and were nearly the same fraction of critical volumes. His effort to evaluate c as an

additive of atomic parameters was not successful.

I propose a modification of Batschinski's formulation as follows. Fluidity depends on the the ratio of free volume, $V - V_0$, to intrinsic volume, V_0, the molal volume at which fluidity is zero. I prefer molal volume to specific volume for conceptual reasoning. By analogy with Batschinski's procedure I write

$$\eta = C/(V - V_0) \qquad (1)$$

I plot measured values of fluidity, $1/\eta$, against molal volume, V. All the simple liquids I have examined, added to the scores Batschinski investigated, give straight lines that yield values of V_0 at the intercept, where $1/\eta = 0$. The slopes of the lines give values of C. Typical cases are illustrated in Figs. 1 and 2 showing the variation of viscosity with temperature of C_3H_8, C_6H_6, and CCl_4. Figure 3 is a plot of fluidity against V for C_6H_6 and CCl_4. Values of V_0 and C are given in Table 1 for these and other liquids, determined in this way. Batschinski, by multiplying his values of specific volume at zero fluidity by molecular weight, obtained values of V_0 which quite agree with mine for the same liquids. He discovered that the ratios of his values of V_0 to the critical volumes for nonassociated liquids vary but little from 0.31. In the cases of CCl_4, CS_2, and P_4, which freeze to solids that retain rotational energy, their solid molal volumes, V_s, are 87.9, 49.0, and 67.7 cm³, respectively, close to their values from fluidity, given in Table 1. The values of viscosity for most of these liquids are from the handbook (5). The data for P_4 are from Powell, Gilman, and Hildebrand (6); those for xylene are from Smith and Hildebrand (7).

It is obvious from the small scatter

of the points from the lines in Fig. 3, which yielded the parameters V_0 and C, that the viscosities of the liquids could be recalculated accurately by Eq. 1 with the parameters in Table 1.

I have modified Eq. 1 so as to recognize the fact that fluidity must depend upon relative expansion, $(V - V_0)/V_0$:

$$\frac{1}{\eta} = B \frac{V - V_0}{V_0} \qquad (2)$$

The parameter $B = V_0/C$. Values of the three parameters, V_0, C, and B, are listed in Table 1. It is striking that the values of B for such different liquids as C_3H_8 and C_7H_{16} are the same within the inaccuracies of the data. This may be explained by the orienta-

Table 1. Parameters of Eqs. 1 and 2.

Molecule	V_0 (cm³/mole)	C	B
C_3H_8	61	3.27	18.6
C_5H_{12}	92	5.10	18.5
C_6H_{14}	111	6.18	18.0
C_7H_{16}	130	6.88	18.8
C_8H_{18}	147	8.58	17.1
C_9H_{20}	165	10.0	16.5
$C_{10}H_{22}$	183	12.2	15.2
$C_{12}H_{26}$	217	15.4	14.1
C_6H_6	82.0	4.44	18.5
p-$C_6H_4(CH_3)_2$	114.5	6.24	18.4
CCl	88.3	5.06	17.4
P_4	68.6	4.22	16.3
CS_2	50.5	3.59	14.0
$CHCl_3$	70	5.00	14.0
Hg	14.10	1.12	12.6

tion of molecules with their long axes in the direction of flow. The progressive decrease in B among the higher members may be the result of increasing flexibility. I can report that the values of V_0 for the normal paraffins are accurately linear with molecular weight, a fact that can serve in interpolating missing values.

It is instructive to consider the magnitude of expansion as revealed by the values of fluidity. Table 2 gives values at 20° and 40°C calculated by Eq. 2 from values of $1/B\eta$. We see that at 20°C CCl_4 has to be only 6 percent expanded over its intrinsic volume in order to have the rather low viscosity of 0.97 cp. Again, C_6H_6 has a viscosity of 0.65 cp when only 8 percent expanded. Such small free volumes correspond to mean free paths of only a few percent of molecular diameters. These figures are in harmony with the calculations for hard sphere fluids of Alder and Einwohner (8) who reported that the probability of a mean free path as long as a molecular diameter in a liquid for which $V/V_0 = 1.6$ is 4×10^{-8}.

The fact that a liquid can be so fluid although expanded so little over its intrinsic volume is evidence of the unreality of some of the concepts that have been used by writers on transport theory, concepts such as "trajectories" between separable "collisions," some of which are "hard" and others "soft," and "cages" in which a molecule is "oscillating" with a definite frequency while awaiting an access of "activation energy" sufficient to enable it to break through a "barrier" into a "hole" awaiting it at a distance of exactly one diameter.

Applying to diffusivity the results of our study of fluidity, we can expect, first, that both begin at the same temperature and molal volume as linear functions of $(V - V_0)/V_0$. Experimental evidence in the case of diffusivity is small in volume but good in substance and quality. I begin with recent determinations of the self-diffusion coefficients of C_6H_6 and CCl_4 between 15° and 50°C by Collings and Mills (9) (Fig. 4). The lines are quite straight and extrapolate to $-35°$ and $-25°C$, respectively. C_6H_6 freezes to a laminar solid at 6°C, but CCl_4 freezes to a solid with rotating molecules and a molal volume of 87.9 cm³ that is close to 88.3 cm³, the value derived from viscosity. V_0 for C_6H_6 was found to be 82.0 cm³ from viscosity; this is its critical molal volume, 256 cm³, multiplied by 0.32, the factor that holds for a number of liquids.

Watts, Alder, and Hildebrand (10) had earlier measured the self-diffusion of CCl_4, obtaining values about 0.1 higher but parallel to those of Collings and Mills.

The greater slope of the line for C_6H_6 in Fig. 4 may be explained without a complete analysis of the problem. Molecules of C_6H_6 in free space move much faster than those of CCl_4. From the equation $1/2\ mv^2 = 3/2\ kT$, where m denotes molecular mass, v is the velocity in free space, and k is the Boltzmann constant, it follows that at the same temperature $v_B/v_A = (m_A/m_B)^{1/2}$; therefore, letting m_B stand for benzene and m_A for CCl_4, since $(m_A/m_B)^{1/2} = (154/78)^{1/2}$, $v_B/v_A =$

Table 2. Values of relative free volume, $(V - V_0)/V_0$, at 20° and 40°C, calculated from $1/\eta B$.

Liquid	T (°C)	η (centi-poise)	$\dfrac{V - V_0}{V_0}$
CCl_4	20	0.97	0.059
	40	.74	.078
C_6H_6	20	.650	.080.
	40	.492	.112
CS_2	20	.366	.184
	40	.349	.194

1.41. The ratio of the slopes of the lines in Fig. 4 is 1.40

Self-diffusion is actually a misnomer; it is determined by introducing a small concentration of molecules tagged so

that they can be recognized, but in a way that does not measurably alter their diffusivity. Under this restriction they form an ideal solution; there is no enthalpy of dilution and the free energy of dilution is T times the entropy of dilution. This kind of free energy is purely statistical; it is not a vector force acting upon individual molecules, impelling them "downstream" against a resistance.

Passing on next to the diffusivity of a dilute solute, let us consider the diffusivity of iodine in CCl_4, determined in 1953 by Haycock, Alder, and Hildebrand (11). Their points have been added to Fig. 4. Because the solvent controls the motions of isolated solute molecules, the straight line through the points extrapolates to the same temperature as the line for the solvent. It is moderately displaced upward because the cross section of iodine molecules is less than that of CCl_4.

The line for dilute CCl_4 in C_6H_6 lies below the line for the solvent by almost exactly the ratio of their slopes, 0.87, compared with the ratio of the ⅔ power of their partial molal volumes, 0.89.

That the line of D plotted against T for a solute starts at the temperature of zero fluidity of the solvent is strikingly illustrated by the data obtained by Ross and Hildebrand (12) for diffusion of gases in CCl_4. They found that the ratios of diffusivity at 25°C to those at 0°C were 1.41 for CH_4 and CF_4, and 1.48 for N_2. The ratio of $(V - V_0)/V_0$ for CCl_4 at 25°C to that at 0°C is 1.47. So much for the temperature coefficient of solute diffusivity.

They found also that the diffusion constants of various solutes and that of isotopic CCl_4 are inversely proportional to their molecular cross sections. Using "best" values of molecular diameters,

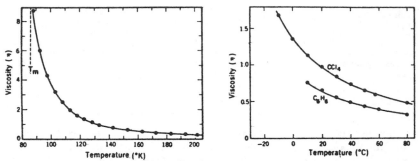

Fig. 1 (left). Variation of viscosity (η) of C_3H_8 with temperature; viscosity is in centipoise. Fig. 2 (right). Variation of viscosity (η) of C_6H_6 and CCl_4 with temperature.

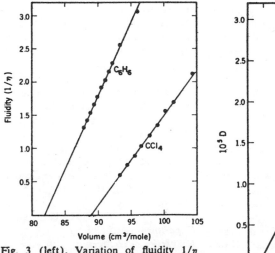
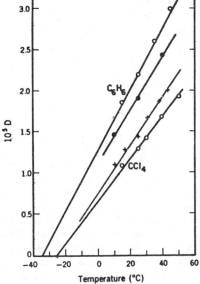

Fig. 3 (left). Variation of fluidity $1/\eta$ with molal volume of C_6H_6 and CCl_4. Fig. 4 (right). Variation of diffusivity (D) (in cm²/sec) with temperature for C_6H_6 and CCl_4 ($\bigcirc$), CCl_4 in C_6H_6 ($\bullet$), and I_2 in CCl_4 ($+$).

σ, they obtained the following relative values of $D\sigma^2$ for solutes and solvent: 47 for CCl_4, 44 for CF_4, 43 for CH_4, and 46 for N_2. The product became 66 for D_2 and 86 for H_2 from the entrance of a quantum effect. This increases linearly with the square of de Boer's quantum mechanical parameter

—excess of zero point energy over classical translational energy—to include ^4He (*13*) and ^3He (*14*). The diffusivity of the nonquantum gases in CCl_4 can be calculated with remarkable approximation from the cross section and the self-diffusivity of CCl_4. Relative values of molecular cross

Table 3. Observed diffusivity of gases in CCl_4 at 25°C compared with values calculated from the self-diffusion coefficient of CCl_4 and ratios of molecular cross sections.

Quantity	CCl_4	CF_4	CH_4	N_2	Ar
V_c (cm³/mole)	276	145	98.6	90.0	74.5
$V_c^{2/3}$	42.5	27.5	21.3	20.1	17.7
10^5D(obs) (cm²/sec)	1.41	2.04	2.78	3.42	3.63
10^5D(calc)(cm²/sec)	——	2.18	2.81	2.97	3.38

sections in fluids can be obtained from critical volumes (V_c). Table 3 gives values of V_c and $V_c^{2/3}$ for CCl_4 and four gases. The diffusivity of these gases in CCl_4 at 25°C has been calculated from the coefficient of self-diffusion of CCl_4, which I take as 1.41 $\times 10^{-5}$ (10) multiplied by the ratio of $V_c^{2/3}$ for CCl_4 to $V_c^{2/3}$ for the gas. This procedure yields the values of $10^5 D$ (calc), which can be compared with the experimental values $10^5 D$ (obs), in the table. In view of the difficulty of making precise determinations of diffusivity and the long extrapolations involved, the agreement is remarkable.

Attempts to calculate absolute values of diffusivity have been made by starting from Stokes' law for a particle settling under the pull of gravity, and extrapolating over the long path to a molecule participating with its neighbors in aimless thermal motions that are never as long as the molecular diameter. Individual molecules are not impelled by a vector force that can serve to measure a "coefficient of friction." Any such coefficient is fictitious.

If molecules were hard spheres, instead of electron clouds with imbedded nuclei, and were sufficiently far apart to justify speaking of binary collisions with linear free paths between them, the probable distance they could be expected to wander from their initial positions could be computed by the formula for a "random walk." But polyatomic molecules that move less than 10 percent of their diameter require a more sophisticated mathematical formulation. They are in a continual state of soft, slow collision, with constant exchange between translational and internal energy. The random walk in this case is a slow, tipsy reel, without sudden changes of direction. The mathematical problem involved is similar to that of calculating the probable distance between the two ends of a string, say 100 feet long, after gathering it rapidly into a tight ball. Since I am interested only in getting transport theory on a course upon which it can be expected to progress toward a solution, I make no effort, at least at present, to try to solve the problem of the ball of string. To start something that tempts others to carry on can be more rewarding than something one can finish alone.

JOEL H. HILDEBRAND
Department of Chemistry,
University of California,
Berkeley 94720

References and Notes

1. R. B. Bird, W. E. Stewart, E. N. Lightfoot, *Transport Phenomena* (Wiley, New York, 1960), pp. 28–29.
2. P. A. Johnson and A. L. Babb, "Liquid diffusion of non-electrolytes," *Chem. Rev.* 56, 387 (1956).
3. I. Prigogine, Ed., *Proceedings of the International Symposium on Transport Processes in Statistical Mechanics* (Interscience, New York, 1958).
4. A. J. Batschinski, *Z. Physik. Chem.* 84, 643 (1913).
5. B. H. Billings and D. E. Gray, Eds., *American Institute of Physics Handbook* (McGraw-Hill, New York, 1963), pp. 2–166.

6. R. E. Powell, T. S. Gilman, J. H. Hildebrand, *J. Amer. Chem. Soc.* **73**, 2525 (1951). (Data for P_4.)

7. E. B. Smith and J. H. Hildebrand, *J. Chem. Phys.*, **40**, 909 (1964).

8. B. J. Alder and T. Einwohner. *ibid.* **43**, 3399 (1965).

9. A. F. Collings and R. Mills, *Trans. Faraday Soc.* **66**, 2761 (1970).

10. H. Watts, B. J. Alder, J. H. Hildebrand, *J. Chem. Phys.* **23**, 659 (1955).

11. E. W. Haycock, B. J. Alder, J. H. Hildebrand, *ibid.* **21**, 1601 (1953).

12. M. Ross and J. H. Hildebrand, *ibid.* **40**, 2397 (1964).

13. K. Nakanishi, E. M. Voigt, J. H. Hildebrand, *ibid.* **42**, 1960 (1965).

14. R. J. Powell and J. H. Hildebrand, *ibid.*, in press.

15. This report was presented as the Gilbert Newton Lewis Memorial Lecture at the University of California, Berkeley, on 25 October 1971

10 September 1971

2

Diffusivity of ^{3}He, ^{4}He, H$_2$, D$_2$, Ne, CH$_4$, Ar, Kr, and CF$_4$ in (C$_4$F$_9$)$_3$N

R. J. Powell* and J. H. Hildebrand

Department of Chemistry, University of California, Berkeley, California 94720

(Received 14 June 1971)

An earlier paper from this laboratory reported that diffusion coefficients of gases in CCl$_4$ multiplied by their molecular cross sections are the same for gases heavier than Ne, but that they increase linearly from D$_2$ to ^{4}He with the square of the de Boer quantum parameter Λ*. The present study of gases in (C$_4$F$_9$)$_3$N includes ^{3}He. For relative cross sections we use, instead of uncertain values of σ^2, the $\frac{2}{3}$ power of the critical volumes, V_c, accurately known from critical densities. Measured values conform to the equation $10^2 DV_c{}^{2/3} = 70(1+1.94\Lambda^{*2})$. Diffusion coefficients in CCl$_4$, multiplied by $V_c{}^{2/3}$ yield $10^2 DV_c{}^{2/3} = 70(1+3.00\Lambda^{*2})$. The relative steepness of the two quantum lines is $3.00 \div 1.94 = 1.60$, exactly the same as the ratio of the internal pressures of the two liquids, 3350/2160.

This is the fifth paper in a series designed to throw light upon the process of diffusion in liquids. The first two[1,2] reported diffusion coefficients of I$_2$ and isotopic CCl$_4$ in CCl$_4$ over ranges of temperature of some 40° and at 1, 65, and 200 atm. These yielded values for both $(\partial D/\partial T)_p$ and $(\partial D/\partial T)_v$. They indicated that diffusing molecules virtually never make jumps as long as molecular diameters.

The third paper[3] reported values of diffusivity in CCl$_4$ at 0° and 25° of H$_2$, D$_2$, Ne, Ar, CH$_4$, CF$_4$, and isotopic CCl$_4$. We found that values of the product of the diffusion coefficients by the squares of the molecular diameters, σ, i.e., $10^{21} D\sigma^2$, lay between 43 and 47 at

Reprinted from:

THE JOURNAL OF CHEMICAL PHYSICS

VOLUME 55, NUMBER 10   15 NOVEMBER 1971

25°C for all the above gases except D_2 and H_2, for which they were 66 and 86, respectively. Data for σ were "best values" from those collected by Hirschfelder, Curtiss, and Bird.[4] This study showed also that attractive forces play little or no role.

The next study[5] was extended to include ^4He, O_2, and SF_6. The value of $10^{21}D\sigma^2$ was 45 for O_2, 43 for SF_6, and $\sim$132 for ^4He. The values for the quantum gases gave straight lines when plotted against squares of the parameter Λ * introduced by de Boer and Michels.[6] It represents the ratio of the de Broglie wavelength to σ.

The purpose of this study was to extend the series of quantum gases to include ^3He. Colleague Alder obtained a supply of this gas from the Lawrence Radiation Laboratory sufficient for determining its solubility and its coefficient of diffusion in $(C_4F_9)_3N$. We changed to this solvent because of its high density, which ensures absence of convection, and its low vapor pressure, 0.3 torr at 25°C, favorable to accuracy in the measurement of the amount of gas passing through the diaphragm. We received an adequate supply of this liquid from the Minnesota Mining and Manufacturing Company, for which we are very grateful.

EXPERIMENTAL

The apparatus is similar to the one described in Ref. 5. The diaphragm is a thicker slice, 2.27 cm, of the cylinder from which the former one was cut. It consists of 2962 pieces of stainless-steel hypodermic tubing embedded parallel in solder. The total diffusion cross section is 5.65 cm². The gas space in the glass dome above the diaphragm has been reduced from 200 to 55 cc by a block of stainless steel shaped to fit loosely in the dome. It stands upon three short legs on the outer rim of the diaphragm. The volume of the vessel below the diaphragm is 212 cc. The amount of liquid was just sufficient to cover the upper surface of the diaphragm to a depth of ~0.5 mm.

Diffusivity was determined as described in Ref. 5 by measuring the rate of decrease in the volume of gas above the diaphragm, maintained at constant pressure, by balancing with a manometer whose other arm was connected to a 10-liter flask of air thermostatted in the bath containing the cell. Temperature fluctuation of 0.01° in the thermostat corresponded to ΔP of 0.02 and ΔV of 0.001 cc.

When gas is introduced into the apparatus, it first saturates the thin layer of solvent above the diaphragm, then establishes a uniform concentration gradient within the diaphragm, after which the amount of gas disappearing into the liquid is linear with time. This steady state is achieved after about 12 h. The dissolved gas is kept distributed uniformly in the lower vessel by means of a magnetic stirrer that revolved 5 times every 3 min. A small correction was made for its slowly increasing concentration in the lower vessel.

The $(C_4F_9)_3N$ had been distilled through a column of many plates; the portion used boiled between 174.7 and 175.5°C. It gave a single peak in vapor-phase chromatography on several different columns, and dissolved iodine showed no uv peak such as would be expected from a hydrocarbon impurity. Its density conformed to the expression: $1.8801[1-0.0022(t-25°)]$. It was thoroughly degassed before use by the standard technique of freezing and melting. The gases other than ^{3}He were manufacturer's "research grade."

RESULTS

Measured values of the diffusion coefficients of the dissolved gases are given in Table I. Concentrations of gas above the diaphragm were calculated from their mole fractions, x, at 25°C in Column 1, taken from a paper to be published in the Journal of Chemical and Engineering Data. The table shows the good agreement obtained between duplicate determinations of diffusivity.

Data reported in Ref. 5 for diffusion of certain gases in $(C_4F_9)_3N$ were made with a diaphragm whose dimensions were known less accurately and without accurate figures for the solubilities of the gases. Furthermore, the density of the liquid then used was 1.872 at 25°C, compared with 1.880 of the present liquid.

TABLE I. Solubility of gas in liquid above diaphragm; $x \equiv$ mole fraction; diffusion coefficients, D, in square centimeters/second; critical molal volumes of gases, V_c in cubic centimeters, values of $10^2 DV_c^{2/3}$, and quantum mechanical parameter, Λ^*.

	$10^4 x$	$10^4 D$	V_c	$10^2 DV_c^{2/3}$	Λ^*
^{3}He	11.02	14.23, 14.41	72.8	248	3.08
^{4}He	11.67	13.76, 13.81	57.8	205	2.67
H_2	15.52	8.21, 8.30	64.8	133	1.73
D_2	16.03	6.65, 6.61	59.9	101	1.22
Ne	16.79	6.34, 6.37	41.7	76.3	0.59
CH_4	68.83	3.14, 3.21	98.6	72.0	0.24
Ar	61.05	3.83, 3.74	74.5	67.0	0.10
Kr	111.5	2.93, 2.92	107.5	66.5	$\sim$0
CF_4	148.1	2.15, 2.15	159.	63.2	$\sim$0

DISCUSSION

The values of σ used in the previous papers to give molecular cross sections are very unsatisfactory for our purpose because they are artifacts that depend upon the particular formula used for pair potentials and yield scattered values.[7] For this study we express relative molecular cross sections by the $\frac{2}{3}$ power of critical volumes, values for which have been determined directly from diameters of liquid–vapor density loops. Critical densities are given in a book edited by Cook,[8] and in the *American Institute of Physics Handbook.*[9]

The apparent quantum effect reported previously seems confirmed by the increase of $10^2 DV_c^{2/3}$ from 205 for ^{4}He to 248 for ^{3}He, 4 times its value for CF$_4$. Its values for the four quantum gases conform to the expression: $10^2 DV_c^{2/3} = 70(1 + 1.94 \Lambda^{*2})$. The earlier data for diffusivity in CCl$_4$, recalculated to the same basis, give $10^2 DV_c^{2/3} = 70(1 + 3.00 \Lambda^{*2})$. It is striking

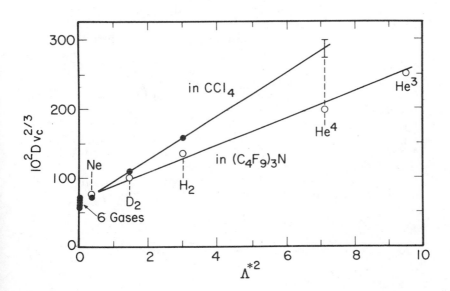

that the relative steepness of the two lines, $3.00/1.94 = 1.60$, is the same as the ratio of the internal pressures of the two liquids, $3350/2160 = 1.60$.

It is a curious fact that it is the difference in values of V_c that makes the larger contribution to the difference between $DV_c^{2/3}$ for ^{3}He and ^{4}He, whereas different values of D make the larger contribution to the difference between H_2 and D_2. Heisenberg uncertainty appears to be differently distributed in the two isotopic pairs.

ACKNOWLEDGMENTS

We thank the 3M Company for the perfluoro-tributylamine and Berni J. Alder for helpful discussions and good offices in obtaining the supply of ^{3}He.

* Present address: Department of Chemistry, The University, Southampton, England.
[1] E. W. Haycock, B. J. Alder, and J. H. Hildebrand, J. Chem. Phys. **21**, 1601 (1953).
[2] H. Watts, B. J. Alder, and J. H. Hildebrand, J. Chem. Phys. **23**, 659 (1955).
[3] M. Ross and J. H. Hildebrand, J. Chem. Phys. **40**, 2397 (1964).
[4] J. O. Hirschfelder, C. F. Curtiss, and R. B. Bird, *Molecular Theory of Gases in Liquids* (Wiley, New York, 1964).
[5] K. Nakanishi, E. M. Voigt, and J. H. Hildebrand, J. Chem. Phys. **42**, 1860 (1965).
[6] J. de Boer and A. Michels, Physica **5**, 945 (1938), also Ref. 4.
[7] J. H. Hildebrand, Proc. Natl. Acad. Sci. U.S. **64**, 1331 (1969).
[8] *Argon, Helium and the Rare Gases*, edited by Gerhard A. Cook, (Interscience, New York, 1961), Vol. 1.
[9] *American Institute of Physics Handbook*, edited by D. E. Gray, *et al.* (McGraw-Hill, New York, 1957), 2nd ed.

3

Fluidity: A General Theory

J. H. HILDEBRAND AND R. H. LAMOREAUX

Department of Chemistry, University of California,
Berkeley, Calif. 94720

Contributed by J. H. Hildebrand, August 5, 1972

ABSTRACT   The equation $\phi = B(V - V_0)/V_0$, which reproduces the fluidity of simple liquids accurately over ranges between freezing and boiling points, is here shown to hold to pressures of at least 500 atm, and nearly to critical volumes. Fluidity can vary continuously above the critical region into that of compressed gas, where the parameter B becomes a function of temperature.

The parameter V_0 is a "corresponding states" fraction of the critical molal volume. It is identical with the molal volume of the solid only in cases where molecules are free to rotate as they do in the liquid.

Parameter B is a measure of the extent to which the external momentum that produces viscous flow is absorbed by the molecules of the liquid. Such damping can result from molecular mass, e.g., Ne, Ar, Kr; flexibility, normal alkanes; or rotational inertia, $SiBr_4$ against $SiCl_4$.

In a previous paper (1) on *Motions of Molecules in Liquids: Viscosity and Diffusivity,* various evidence was presented showing that the fluidity of a simple liquid, ϕ, at atmospheric pres-

Reprinted from
Proc. Nat. Acad. Sci. USA
Vol. 69, No. 11, pp. 3428–3431, November 1972

sure and at any temperature, is proportional to the fractional excess of its molal volume, V, over the molal volume, V_0, at which the molecules are so closely crowded as to prevent viscous flow while still retaining rotational freedom. At volumes less than V_0 only plastic flow is possible. This concept is expressed formally by the equation:

$$\phi = B \frac{V - V_0}{V_0}. \qquad [1]$$

The unit of viscosity is the "poise." It can be defined in terms of liquid held between a fixed and a moving plate, 1 cm apart; the viscosity is the force in dynes divided by the area of the moving plate that must be applied to it in order to maintain a steady velocity of 1 cm/sec. The moving plate transmits momentum to the molecules of the liquid. Fluidity, in terms of this arrangement, can be thought of as the velocity of the moving plate, in cm/sec, when the force applied is 1 dyne/cm² of area. The viscosity of ordinary liquids at ambient temperatures is of the order of 10^{-2} poise; hence, it is customary to use the centipoise as the unit. The unit of fluidity is thus a reciprocal centipoise, cP^{-1}.

The value of V_0 for a liquid is obtained, as explained in ref. 1, by plotting experimental values of ϕ at different temperatures against the values of V, the molal volumes of the liquid, at the same respective temperatures. The scores of simple liquids that we have examined in this way all yield straight lines over ranges of temperature from 0° up nearly to boiling points. Extrapolated to $\phi = 0$, they yield values of V_0. The slope of a line is $\phi/(V - V_0)$, which is B/V_0. In the following pages we show that this equation accounts correctly for the fluidity of simple liquids of various types over large ranges of pressure, volume, and temperature. We explain also that V_0 and B are definite physical quantities, not simply adjustable parameters.

Alkanes

We first apply these concepts to alkanes. Values of V_0 and B, determined from values of viscosity and density found in the literature, are given in Table 1. The values of V_0 for normal alkanes increase linearly with molecular weights. They are also "corresponding states" fractions, 0.300, within limits of error, of critical mo!al volumes, as far as these are available, as seen in Table 2. One of us (2) recently called attention to the fact that the mean of ratios of molal volumes of 11 simple liquids at their boiling points, V_b, to their critical volumes is 0.377. Accordingly, values of V_0 can be estimated with only moderate uncertainty from $0.796\ V_b$.

Values of B for normal, branched, and cycloalkanes are plotted in Fig. 1 against the number of carbon atoms in the compound. The most striking feature of such a plot is the straight 'ine for normal alkanes from propane to the longest chains for which data appear trustworthy. We have omitted some badly scattered points for chain lengths above 20, where

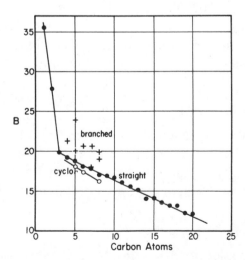

Fig. 1. Variation of the fluidity parameters of alkanes, B, reciprocal centipoises, with number of carbon atoms and molecular structure.

prediction by linear extrapolation would appear to be much more reliable than published data in this region.

One sees, next, that B-values for cycloalkanes are a little smaller than those of the corresponding normal alkanes, and that those of branched chains are a little larger. Especially noteworthy is the series of the pentanes: *cyclo-*, *normal-*, *iso-*, *neo-*. Although the major factor determining fluidity is the crowding of the molecules, expressed by the ratio $(V - V_0)/V_0$, the absorption of energy during molecular collisions involved in the transfer of momentum, represented by B, can be very significant. It is to be expected that molecules with rotational inertia would lose more energy during collision than molecules with none. This effect may account for the steep drop in B values from CH_4 to C_2H_6 to C_3H_8. From then on, the normal alkanes can absorb energy by bending, a loss that is proportional to the length of the chain. We present below further evidence on the relation of B-value to molecular composition and structure.

We emphasize that Eq. 1, with the parameters V_0 and B in Table 1, makes it possible to calculate viscosities of normal alkanes from C_3H_8 to chain lengths of 20 or more carbon atoms with accuracy quite equal to that of measured values, and that the role of temperature is simply to determine the magnitude of $(V - V_0)$.

Effect of pressure on fluidity

The viscosity of propane has been determined at pressures up to 544 atm, as shown in Table 3. The constancy of the ratio $\phi/(V - V_0)$ in the last column shows that Eq. 1 accounts excellently for the effect of high pressures upon fluidity.

Tetrahalides and elements

The four tetrachlorides in Table 4 have B values that decrease sharply from CCl_4 to $SnCl_4$ with increase in size, tetrahedral shape, and moments of rotation that increasingly resist the momentum of viscous flow. The increase in mass from $SnCl_4$

TABLE 1.   *Values of V_0, c^3/mol and B, reciprocal centipoise*

	V_0	B	Reference
CH_4	32.0	35.5	a
C_2H_6	49.6	25.5	a
C_3H_8	61.0	19.8	a
C_4H_{10}	77.7	19.2	a
C_5H_{12}	94.0	18.7	a
C_6H_{14}	111.0	18.0	a
C_7H_{16}	129.1	17.7	a
C_8H_{18}	146.4	17.0	a
C_9H_{20}	165	16.9	a
$C_{10}H_{22}$	183	16.6	a
$C_{11}H_{24}$	201	16.0	a
$C_{12}H_{26}$	217	15.5	a
$C_{13}H_{28}$	238	15.0	a
$C_{14}H_{30}$	253	14.0	a
$C_{15}H_{32}$	273	14.0	a
$C_{16}H_{34}$	291	13.5	a
$C_{17}H_{36}$	309	13.1	a
$C_{18}H_{38}$	329	13.1	a
$C_{19}H_{40}$	345	12.2	a
$C_{20}H_{42}$	365	12.2	a
$C_{24}H_{50}$	434	9.8	b
$C_{35}H_{72}$	622	6.0	b
c-C_5H_{10}	83.3	17.6	a, c
c-C_6H_{12}	103.2	17.2	a, d
c-C_8H_{16}	130.7	12.6	e
2-$CH_3C_3H_7$	82.2	21.3	a
$2,2$-$(CH_3)_2C_3H_6$	104.4	23.9	a
2-$CH_3C_4H_9$	98.2	20.0	a, f
2-$CH_3C_5H_{11}$	113.5	20.7	f, g
2-$CH_3C_6H_{13}$	131.0	20.7	a, f
$2,2,4(CH_3)_3C_5H_9$	150.5	19.9	a, h, i
2-$CH_3C_7H_{15}$	148.0	19.0	c
2-$CH_3C_{10}H_{21}$	201	15.9	j

a.   Rossini, F. D. *et al.* (1971) *American Petroleum Inst. Research Project No. 44.* Carnegie Press, Pittsburgh, Pa.

b. Nederbragt, G. W. & Boelhouwer, J. W. M. (1947) *Physica*
    13, 305–318.
c. Geist, J. M. & Cannon, M. R. (1946) *Ind. Eng. Chem. Anal.
    Ed.* **18**, 611–613.
d. Friend J. N. & Hargreaves, W. D. (1944) *Phil. Mag.* **35**,
    57–64.
e. Kuss, E. *Z. Angew. Physik* (1955) **7**, 372–378.
f. Thorpe, E. & Rodger, J. W. (8974) *Phil. Trans. Roy. Soc.
    Lond.* **A185**, 397–710.
g. Bridgman, P. W. (1931/32) *Proc. Amer. Acad.* **66**, 185–233.
h. Maman, A. (1938) *Comptes Rend.* **207**, 1401–1402.
i. Evans, E. B. (1938) *J. Inst. Petroleum Techn.* **24**, 48–53.
j. Terres, E. & Brinkmann, L. *et al.* (1959) *Brennstoffchemie* **40**,
    279–280.

TABLE 2.   *"Corresponding States" ratio of V_0 to critical
volume for normal alkanes c^3/mol*

	V_c	V_0	(V_0/V_c)
CH_4	99.4	32.0	0.322
C_2H_6	148.1	44.6	0.301
C_3H_8	203.2	61.0	0.300
C_4H_{10}	254.9	77.7	0.305
C_5H_{12}	304.4	94.0	0.309
C_6H_4	369.9	111.0	0.300
C_7H_{16}	431.9	129.1	0.299
C_8H_{18}	492.4	146.4	0.297

to $SiBr_4$, which have the same V_0, causes ϕ to drop from 9.0
to 5.5. The decrease in fluidity from 45.2 for Ne to 12.3 for
Kr, with Ar_2 between, must result from increasing atomic
mass.

It is not possible, because of lack of independent data, to
calculate quantitatively the effect upon fluidity of molecular
flexibility, softness, and inertia of rotation. In fact, fluidity of
liquids itself offers the most significant information to be had
about molecular repulsions. The difference between $B = 45.0$

for neon and 22.7 for argon clearly shows how naive it is to use the same inverse 12th-power to express their repulsive potential.

Distinction between V_0 and V_s, the molal volume of the solid

Others have sought to identify V_0 with V_s. This is possible only for substances whose molecules retain rotational energy in the solid state, as do those of CCl_4; most species of poly-

TABLE 3. *Values of $\phi/(V - V_0)$ for propane at 311°K and different pressures[a,b]*

P, atm	ϕ, cP^{-1}	V, c^3/mol	$\phi/(V - V_0)$
13.60	11.79	93.3	0.365
68.05	10.46	90.0	0.361
136.1	9.31	87.0	0.358
272.2	7.84	83.0	0.356
544.4	6.21	78.3	0.359

a. Reamer, H. H. Sage, B. H. & Lacey, W. N. (1949) *Ind. Eng. Chem.* **41**, 482–484.
b. Starling, K. E. Eakin, B. E. & Ellington, R. T. (1960) *AIChE J.* **6**, 438–442.

atomic molecules are able to crystallize in a less symmetrical, nonrotating, denser form of lower energy. In such cases, V_0 can be reached only by undercooling the liquid. *Para*-xylene (3), for example, freezes at 13.2° from $V = 122.5$ cm^3 to $V_s = 101.9$ cm^3, much smaller than the volume required for free rotation in the liquid, where $V_0 = 114.5$ cm^3. *Meta*-xylene whose density, viscosity, and heat capacity are nearly the same as those of p-xylene in their common liquid range above 13.2°, cannot freeze into a comparable closely packed solid form, and remains liquid to $-47°$. Its value of V_0 is 113.5 cm^3.

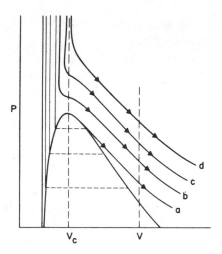

Fig. 2.   Van der Waals paths from liquid to gas along different isotherms.

Cogwheel molecules

Magill and Ubbelohde (4) have prepared substances, such as tridiphenyl methane, whose molecules have three long arms. They found their viscosities, plotted as $\log \eta$ against $1/T$, gave lines that acquired curvature above the freezing point, an effect that they attributed to "prefreezing." Plotted in our

TABLE 4.   *Values of V_0 and B for tetrahalides and elements. Units cm^3/mol and cP^{-1}*

	V_0	B	Reference
CCl_4	89.9	13.4	a, b
$SiCl_4$	97.5	12.6	c, d, e
$TiCl_4$	98.3	10.2	c, f
$SnCl_4$	104.7	9.0	c, g
$SiBr_4$	104.9	5.5	c, h
Ne	14.3	45.2	i, j
Ar	24.5	22.7	i, j
Kr	29.0	12.3	i, j

a. Billings, B. H. & Gray, D. E. Eds., *American Institute of Physics Handbook* (McGraw-Hill, New York, 1963).

b. Young, S. (1910) *Sci. Proc. Roy. Dublin Soc.* **12**, 374–443.

c. Lutschinsky, G. P. (1937) *J. Gen. Chem. (USSR)* **7**, 2116–2127.

d. Bowden, S. T. & Morgan, A. R. (1940) *Phil. Mag.* **29**, 367–378.

e. Hölemann, P. (1936) *Z. Physik. Chem.* **B32**, 353–368.

f. Sagawa, T. (1933) *Sci. Rep. Tôhoku Univ.* **22**, 959–971.

g. Ulich, H. Hertel, E. & Nespital, W. (1932) *Z. Physik Chem.* **B17**, 369–379.

h. Nisel'son, L. A. Sokolova, T. D. & Lapidus, I. I. (1967) *Zh. Neorg. Khim.* **12**, 1423–1426.

i. Förster, S. (1963) *Cryogenics* **3**, 176–179.

j. Agrawal, G. M. & Thodos, G. (1971) *Phys. Chem. Fluids* **2**, 135–145.

way as ϕ against V, such substances give lines that curve to the left as ϕ approaches 0. Their arms become entangled as $(V - V_0)$ approaches 0. Fluidity increases more slowly when $(V - V_0)$ is small than when it is large. We have found this bending of the line in the case of $(C_4F_9)_3N$. It surely is not caused by "prefreezing." Hildebrand and Archer (3) showed that *meta*- and *para*-xylenes have nearly the same liquid molal volumes through their common liquid range, and Smith and Hildebrand (5) found almost identical viscosities, with no evidence of "prefreezing" of the *para*-isomer above its melting point, 13.2°.

Liquid metals also obey Eq. **1,** but we are not able to account for their individual B-values.

Fluidity in the transition from liquid to gas

The linearity observed for plots of ϕ against V for simple liquids below their boiling points made us curious to find out how far it would continue at higher temperatures and volumes.

In order to prevent evaporation, it is necessary to increase pressure, after which expansion can continue along different

TABLE 5. *Parameters for C_3H_8 and CO_2. Volumes, c^3/mol*

	T_c	V_c	V_0	B	Reference
C_3H_8	370	203.2	61.0	19.8	a, b, c
CO_2	304	94.0	25.2	11.0	d, e, f, g

a. Rossini, F. *et al.*, *American Petroleum Institute Research Project No. 44* (Carnegie Press, Pittsburgh, 1953).
b. Reamer, H. H. Sage, B. H. & Lacey, W. N. (1949) *Ind. Eng. Chem.* **41**, 482–484.
c. Starling, K. E. Eakin, B. E. & Ellington, R. T. (1960) *AIChE J.* **6**, 438–442.
d. Stakelbeck, H. (1933) *Z. Ges. Kälteind* **40**, 33–40.
e. *Handbook of Chemistry and Physics, 42nd Ed.* (Chemical Rubber Co., Cleveland, 1961), p. 2465.
f. Kestin, J. Whitelaw, J. H. & Zien, T. F. (1964) *Physica* **30**, 161–181.
g. Michels, A. Botzen, A. & Schuurman, W. (1957) *Physica* **23**, 95–102.

isotherms, as depicted in the van der Waals-type plot of Fig. 2, paths *b*, *c*, and *d*. If the pressure applied is not sufficient to surmount the critical region, ϕ can travel discontinuously along path *a*.

Let us see how ϕ varies as V increases along these different paths. Data for propane and carbon dioxide are plotted in Fig. 3. Significant properties for these compounds are given in Table 5. (Note that the B-value of CO_2, with rigid, linear molecules, is much smaller than that of the more-symmetrical C_3H_8.) The lines below the critical region are quite straight, although ϕ increases to 30 reciprocal centipoise, cP^{-1}, ten times its value for liquids below their boiling points. As V is further increased, the fluidity begins to depend strongly upon the temperature chosen. We explain the effect of temperature as follows: at the critical temperature, V_c is only 2.65 V_0, but as volume increases, the molecules acquire translational momentum increasing in proportion to $(V^{1/3} - V_0^{1/3})$, up to

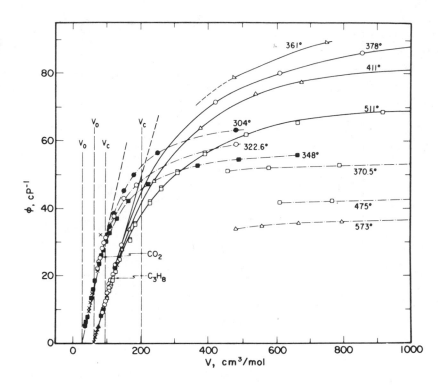

Fig. 3. Variation of fluidity from liquid to gas at different temperatures and pressures.

its kinetic theory value, $(3mkT)^{1/2}$. This random momentum offers resistance to the vector momentum imposed when viscosity is measured; therefore, ϕ varies inversely with $T^{1/2}$, as seen in Fig. 4 for propane. This relation for dilute gases is well known from kinetic theory (4). The departure in the critical region is not unexpected. The difference between a pair of lines in Fig. 3 increases in parallel with increases in $(V^{1/3} - V_0^{1/3})$. We are not ready to reduce this physical description of the process to precise mathematical terms.

NOTE ADDED IN PROOF

Eicher, L. D. & Zwolinski, B. J. (1972) *Science* **177,** 369, stated that Eq. **1** of this paper does not adequately agree with

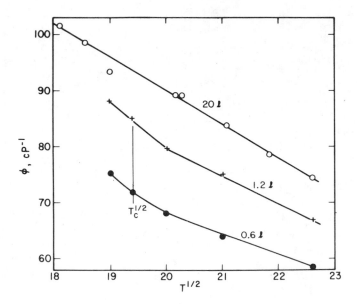

Fig. 4.   Plots of ϕ against $T^{1/2}$ at $V = 0.60$, 1.20, and 20 liters.

data for viscosity of *n*-hexane, *n*-decane, *n*-heptadecane, and *l*-pentanol. They arrive at this conclusion by an analytical application of least squares.

The degree of agreement between theory and experiment is usually revealed far more clearly in a suitable plot than from an analytical application of the method of least squares. In the case of hexane, for example, a plot of ϕ against V of the data referred to in Table 1*a*, whose temperatures range from 70° to −90°, plotted on a 50-cm scale fall on a straight line with scarcely perceptible deviations. Indeed, the values of B and V_o for the scores of liquids given in this paper have all been obtained from such straight-line plots.

This work has been supported by the National Science Foundation and by the Petroleum Research Fund of the American Chemical Society.

1. Hildebrand, J. H. (1971) *Science* **174**, 490–493.
2. Hildebrand, J. H. (1969) *Proc. Nat. Acad. Sci. USA* **64**, 1331–1334.
3. Archer, G. & Hildebrand, J. H. (1961) *Proc. Nat. Acad. Sci. USA* **47**, 1881–1882.
4. Magill, J. H. & Ubbelohde, A. R. (1958) *Trans. Faraday Soc.* **54**, 1811–1821.
5. Smith, E. B. & Hildebrand, J. H. (1964) *J. Chem. Phys.* **40**, 909–910.

4

Fluidity and Liquid Structure

Sir: Hildebrand[1] published a paper in 1971 on "Motions of Molecules in Liquids: Viscosity and Diffusivity" in which he altered an equation for the viscosity of unassociated liquids published by Batschinski[2] in 1913

$$\eta = c/(\nu - \omega) \tag{1}$$

c is a constant for each liquid, ν is specific volume, ω is similar to the van der Waals b. Hildebrand reasoned that fluidity, $\phi = 1/\eta$, should be a linear function of the ratio of intermolecular volume to the volume, V_0, at which, as temperature decreases, molecules become too closely crowded to permit either free flow or self-diffusion; he wrote

$$\phi = B(V - V_0)/V_0 \tag{2}$$

B is a constant whose value depends upon capacity of the molecules to absorb momentum because of their mass, flexibility, or inertia of rotation. Plots of ϕ against V showed straight lines over long ranges of temperature. Extrapolation to $\phi = 0$ gives values of V_0, and slopes give B/V_0. Hildebrand and Lamoreaux[3] later gave values of B and V_0 for scores of liquids; they showed that the equation holds for propane for pressures up to 544 atm, that V_0

---

values are fixed fractions of critical volumes, and that
plotted lines can remain straight nearly to critical temperatures.

Early in 1972 Eicher and Zwolinski[4] published a paper
titled "Limitations of the Batschinski–Hildebrand Shear
Viscosity Equation." On the basis of a "least-squares
analysis" they asserted, "It can be seen that the simple
form of eq 1 or eq 2 will not satisfactorily represent the
experimental data for the four substances, n-hexane, n-
decane, n-heptadecane, or 1-propanol over reasonable
temperature ranges, within experimental uncertainties."

The question thus raised is far more important than
merely one of fitting experimental data; equations of very
different kinds can be tailored to fit the same data by introducing adjustable, nonoperational parameters. The
basic question is whether a very simple equation, based
upon the concept of molecular chaos implicit in the van
der Waals equation, is adequate for dealing with transport
processes in liquids, or whether it is necessary to imagine
the presence of "solid-like" structures. Our objective in
studying viscosity has been to compare the validity of different concepts of the liquid state.

Let us see whether the conclusion that Eicher and
Zwolinski have drawn from their mean squares calculation
is correct. In Figure 1 the data for n-hexane by Giller and
Drickamer[5] are plotted as ϕ against V. We see that the
four upper points, between −60 and 20°, surely "a reasonable range," fall on a straight line. The nine divergent
points at the bottom, close to the freezing point −95.3°,
lie between −90.3 and −98.5°. Eicher and Zwolenski must
have given all points equal weight. In doing this they
overlooked what Giller and Drickamer wrote about these
points.

"There is a small but consistent increase in [free energy
of activation] for each compound near the freezing point,
which would represent the increased activation energy
necessary because of a certain degree of order developing
in the liquid." This sort of divergence in a region of high

viscosity is usual with substances whose molecules are so unsymmetrical that they do not gain full freedom of motion till the liquid has expanded a little more after melting. Magill and Ubbelohde[6] showed that this effect can be increased by using species such as tridiphenylmethane. We found it with $(C_4F_9)_3N$ but never with monatomic molecular species.

The second liquid they offer as evidence against eq 2 is n-heptadecane, from measurements by Doolittle and Peterson.[7] The plot in Figure 2 shows a straight line from ~0.8 to 6 cP^{-1}, and points diverging near the lower end as in Figure 1.

(1) J. H. Hildebrand, *Science*, **174**, 490 (1971).
(2) A. J. Batschinski, *Z. Phys., Chem.*, **84**, 643 (1913).
(3) J. H. Hildebrand and R. H. Lamoreaux, *Proc. Nat. Acad. Sci. U. S.*, **69**, 3428 (1972).
(4) L. D. Eicher and B. J. Zwolinski, *Science*, **177**, 369 (1972).
(5) F. G. Giller and H. G. Drickamer, *Ind. Eng. Chem.*, **41**, 2067 (1949).
(6) J. H. Magill and A. H. Ubbelohde, *Trans. Faraday Soc.*, **54**, 1811 (1958).
(7) A. K. Doolittle and R. H. Peterson, *J. Amer. Chem. Soc.*, **73**, 2145 (1951).

TABLE I: Constancy of $\phi/(V - 184.0)$ at 171° up to High Pressures for n-Decane

P, atm	ϕ, cP^{-1}	V, cm^3	$\phi/(V - V_0)$
13.6	4.725	232.2	0.0098
27.2	4.640	231.0	0.0099
54.4	4.505	229.9	0.0098
68.0	4.264	227.9	0.0097
204.0	3.450	220.4	0.0095
340.0	2.989	214.8	0.0097
408.0	2.801	212.3	0.0099

Their third liquid is n-decane, whose viscosity was measured by Lee and Ellington[8] over ranges of 250° and 408 atm. At 1 atm, the plot is quite like those for n-C_6H_{14} and n-$C_{17}H_{36}$. To test the applicability of eq 2 at different pressures, we give in Table I values of $\phi/(V - V_0)$ at 171° and seven pressures up to 408 atm: $V_0 = 184$ cm³. The virtual constancy of the ratio shows that fluidity is uniquely determined by values of $V - V_0$, irrespective of whether changes in volume result from changes of temperature or of pressure.

Their fourth example is an alcohol, an associated liquid, and therefore not pertinent to the validity of eq 2.

It has been asserted that viscosity is not a function of liquid volume alone, that it decreases with increasing temperature at constant volume. We reply by referring to Figure 3 of ref 3 and the accompanying discussion, where variations of V for C_3H_8 and CO_2 are shown from V_0 to 1000 cm³. We reproduce here in Figure 3 a plot of fluidity against molal volume of propane from V_0 to its critical volume, V_c, using data by Starling, Eakin, and Ellington[9] obtained at 378, 444, and 511°K, and at pressures up to 544 atm.

The points for ϕ from 0 to 5 cP^{-1} are at 1 atm, and below the boiling point, the usual range of viscosity data.

The points for ϕ from 5 to about 18 cP^{-1} were determined at various values of pressure and temperature within the range stated above. It can be seen that values of ϕ over this enormous range depend upon values of V only. As V increases further, however, the single line splits increasingly into three, with fluidity at any value of V largest at the *lowest* temperature; in other words, viscosity *increases* with temperature.

As we explained in ref 3, mean free paths at boiling points are very much shorter than molecular diameters. In the case of propane, $(V_b - V_0)/V_0$ is only 0.23. At its critical point, $V_c = 203.2$ and the fractional expansion is 2.32,

(8) A. L. Lee and R. T. Ellington, *J. Chem. Eng. Data*, **10**, 346 (1965).
(9) K. E. Starling, B. E. Eakin, and R. T. Ellington, *AIChE J.*, **6**, 438 (1960).

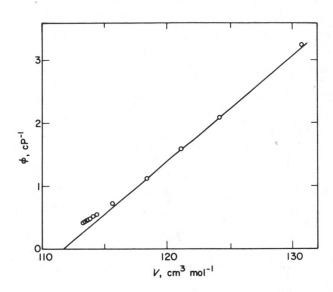

Figure 1. Fluidity against molal volume of *n*-hexane, showing divergence near melting point.

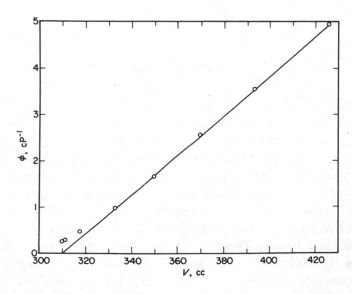

Figure 2. Fluidity against molal volume for *n*-heptadecane.

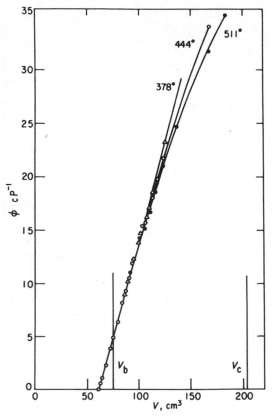

Figure 3. Values of ϕ against V for propane from V_0 to V_c.

where molecules can acquire some momentum between collisions. We have explained that the parameter, B, of eq 2, depends upon the capacity of the species to resist externally imposed momentum by reason of mass, softness, or rotational inertia, or their own thermal momentum, which is, for molecules in free space, $(3mkT)^{1/2}$. This is approached as length of mean free paths increases.

We turn finally to data on five isomeric hexanes measured by Eicher and Zwolinski[10] themselves. Their figures for viscosity and density at different temperatures yield the values of ϕ and V plotted in Figure 4. All points fall

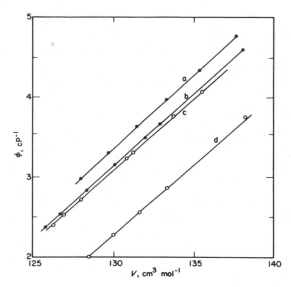

Figure 4. Linearity of $\phi/(V - V_0)$ for five isomeric hexanes.

TABLE II: Evidence for Strict Linearity of the Line for 2-$CH_3C_5H_{11}$ in Figure 4

V, cm³ mol⁻¹	ϕ, cP⁻¹	$\phi/(V - 113.0)$
126.27	2.447	0.1814
126.92	2.528	0.1816
128.03	2.723	0.1812
130.85	3.236	0.1813
131.27	3.308	0.1815
133.82	3.771	0.1811
135.48	4.070	0.1810

on straight lines except one, which is off by only 0.3 cP⁻¹. The excellence of the fit we illustrate in Table II for 2-$CH_3C_5H_{11}$, whose line is the longest of all. The intercept gives $V_0 = 113.0$ cm³. The ratios in the last column show that the line is even straighter than could be inferred from a good plot.

We summarize the foregoing evidence by asserting that (a) the data are accurate, (b) they all conform closely with eq 2, (c) the concept of molecular chaos that served as foundation for it is adequate, and that no highly structured model of the liquid state is required for dealing with transport processes.

Acknowledgment. Acknowledgment is made to the donors of The Petroleum Research Fund, administered by the American Chemical Society, for support of this research.

(10) L. D. Eicher and B. J. Zwolinski, *J. Phys. Chem.*, **76**, 3295 (1972).

Department of Chemistry   J. H. Hildebrand*
University of California   R. H. Lamoreaux
Berkeley, California 94720

Received December 6, 1972

5

Activation Energy: Not Involved in Transport Processes in Liquids

Mean free paths of molecules in simple liquids are very much shorter than their diameters; diffusion occurs by a succession of small displacements, not by leaps through barriers requiring energy of activation. Changes of viscosity and diffusivity with temperature can be accurately and more simply expressed in nonexponential formulas than by plotting their logarithms against reciprocal temperatures.

The method most commonly used for representing the temperature dependence of viscosity and diffusivity of liquids is to plot their logarithms against $1/T$, by analogy with the Arrhenius equation for chemical rate constants, and interpreting the slopes of the lines thus obtained as "activation energy." This designation implies the presence of barriers against freedom of flow, imagined as consisting of some sort of quasi-lattice structure. We wish to call attention to evidence that no activation is involved in these processes, and that their variations with temperature are quite accurately and more simply represented by nonexponential equations.

Reprinted from I&EC FUNDAMENTALS, Vol. 12, Page 387, August 1973
Copyright 1973 by the American Chemical Society and reprinted by permission of the copyright owner

Much of the evidence is summarized in Chapter 3 of "Regular and Related Solutions," by Hildebrand, *et al.* (1970). We invite special attention to a paper by Dymond and Alder (1966), who showed that "calculated values for transport coefficients of the rare gases at temperatures and densities greater than the respective critical ones agree within about 10% with the experimental results in both absolute values and temperature dependence, without involving an activation barrier. In this *a priori* theory, the cross section, that is, the square of the hard-sphere diameter, is determined as a function of temperature from the available equilibrium data, within the framework of the van der Waals theory."

Years of study and experiment with simple liquids have led us to concepts that we describe as follows.

(1)   All molecules participate equally in thermal agitation that produces maximum disorder. It is represented by the distribution functions deduced from X-ray scattering, which yields diffuse rings, quite unlike the spots obtained from crystals. All characteristics that distinguish crystals from liquids disappear upon melting; vacancies that may exist in a crystal become randomly distributed in its liquid as intermolecular space. Consequently, it makes no physical sense to extend to liquids a theory of diffusion in solids that postulates the presence of holes of molecular size. It is not necessary to assume even for diffusion in solids a mechanism that includes activation energy, since realistic values of coefficients of diffusion can be calculated for hard-sphere systems that are without fluctuations in potential energy (Bennett and Alder, 1971).

(2)   No directive force, such as an electric field causing ions to migrate, or gravity causing sedimentation of suspended particles, acts upon the molecules of a nonpolar liquid. They diffuse simply because thermal motions keep them ever on the move. Their mean displacement with time depends (a) upon temperature and (b) upon the ratio of intermolecular volume, V, to the volume, V_0, at which the molecules become too closely crowded to permit either diffusion or bulk flow.

(3)   We have shown (Hildebrand, 1971; Hildebrand and Lamoreaux, 1972) with scores of examples that fluidity, ϕ, the reciprocal of viscosity, over ranges of liquid molal volume from V_0 nearly to the critical molal volume, conforms closely to the equation

$$\phi = B(V - V_0)/V_0 \qquad (1)$$

Values of B depend inversely upon the capacity of molecules to absorb the externally imposed momentum of viscous flow by reason of their mass, flexibility, softness, or inertia of rotation.

The primary effect of temperature, together with pressure, is to determine values of V. Above the critical volume it also alters values of B, as found by Hildebrand and Lamoreaux. As V expands through the critical volume and beyond and paths between collisions become longer, permitting molecular velocities to approach those in free space where momenta are proportional to $T^{1/2}$, fluidity at a specified volume is proportional to $T^{-1/2}$.

A third although minor effect of temperature in this region is a small decrease in effective molecular diameters.

Values of V_0 and B for liquids below their boiling points can easily be obtained by plotting ϕ against V; straight lines are always obtained for simple liquids. The intercept at $\phi = 0$ gives V_0 and the slope gives B/V_0.

Nonspherical molecular shapes may reduce freedom of motion when $V - V_0$ is very small, resulting in some bending away from the straight line near the bottom.

This model of the motions of molecules in liquids applies equally well to diffusion, as outlined by Hildebrand (1971). Dymond (1972) has since shown by molecular dynamics that exact self-diffusion coefficients for hard-sphere systems can be calculated from the ratio V/V_0, where V_0 is the volume of close-packed spheres.

When we published our first paper on diffusion (Haycock, et al., 1953), we too plotted log D against $1/T$, but during the ensuing years we gradually learned more about the liquid state until we now "have a better idea" that we gladly share with others in this paper.

B. J. ALDER

Lawrence Livermore Laboratory
University of California
Livermore, Calif. 94450

J. H. HILDEBRAND*

Department of Chemistry
University of California
Berkeley, Calif. 94720

RECEIVED for review March 21, 1973
ACCEPTED April 3, 1973

Acknowledgment is made to the U. S. Atomic Energy Commission and to the donors of The Petroleum Research Fund, administered by the American Chemical Society, each for partial support of this research.

Literature Cited

Bennett, C., Alder, B. J., *J. Chem. Phys.* **54,** 4796 (1971).
Dymond, J. H., *Trans. Faraday Soc.* **68,** 1789 (1972).
Dymond, J. H., Alder, B. J., *J. Chem. Phys.* **45,** 2061 (1966).
Haycock, E. W., Alder, B. J., Hildebrand, J. H., *J. Chem. Phys.* **21,** 1601 (1953).
Hildebrand, J. H., *Science* **174,** 490 (1971).
Hildebrand, J. H., Lamoreaux, R. H., *Proc. Nat. Acad. Sci. U. S.* **69,** 3428 (1972).
Hildebrand, J. H., Prausnitz, J. M., Scott, R. L., "Regular and Related Solutions," Van Nostrand–Reinhold, New York, N. Y., 1970.

6

VISCOSITY ALONG CONTINUOUS PATHS
BETWEEN LIQUID AND GAS

J.H. HILDEBRAND and R.H. LAMOREAUX

*College of Chemistry, University of California,
Berkeley, Ca. 94720, USA*

Received 8 October 1973

Synopsis

Three sources combine to determine viscosity in the continuous range from dense liquid to expanded gas: (1) the capacity of a species of molecule to absorb the momentum imposed to produce newtonian flow; (2) the ratio of unoccupied to intrinsic volume; (3) opposition to this vector momentum by the random thermal momentum of molecules at the temperature of the fluid.

The first two of these contributions to viscosity are significant at all volumes less than approximately half of the critical molal volume, V_c, and conform to the linear equation for fluidity, $\bar{\phi} = B(V - V_0)/V_0$. Its reciprocal gives their contribution to viscosity,

$$\eta_a = V_0/B(V - V_0).$$

This is independent of temperature.

The contribution of thermal momentum to viscosity we express by

$$\eta_b = [1 - (0.5V_c/V)^{2/3}]\,\eta_0,$$

where η_0 is the limiting dilute-gas viscosity at the stated temperature when $V \gg V_c$.

Hildebrand[1]) published a paper in 1971 in which he altered and amplified a formulation for liquid viscosity published by Batschinski[2]) in 1913 to the form:

$$\phi = B(V - V_0)/V_0, \tag{1}$$

where ϕ is fluidity; V is the molal volume; V_0 is the volume at the temperature where $\phi = 0$; B is a constant of proportionality which is smaller the greater the capacity of the molecules to absorb the external momentum imposed to produce newtonian flow, because of their softness, flexibility, or rotational inertia.

Hildebrand and Lamoreaux[3]) in a paper with the title: "Fluidity: A General

Theory", reported that plots of ϕ against V give lines that continue quite straight under pressure to far above boiling points. In the case of propane, ratios of ϕ to $V - V_0$ remain constant for pressures as high as 544 atm at all volumes less than about half of the critical volume. (See fig. 1.) But when volume is allowed to expand toward the critical volume and beyond, the line separates into isotherms and bends toward horizontal, as illustrated in fig. 2 for propane and carbon dioxide, here reproduced from ref. 4. This was explained qualitatively as the result of the opposition by the random thermal molecular momenta to the directed external momentum applied to cause newtonian flow.

We are now prepared to offer a quantitative treatment of what is shown in fig. 2. We reason in terms of viscosity instead of fluidity, however, since different contributions to viscosity are additive, and we write the reciprocal of eq. (1),

$$\eta_a = V_0/B(V - V_0). \tag{2}$$

This, unlike eq. (1), is not linear, therefore plots of ϕ against V should be used to evaluate V_0 and B. The linearity persists to approximately 0.5 times the critical volume, V_c, at least for the two fluids here to be considered, propane and carbon dioxide.

Beyond about $0.5V_c$, mean distances between molecules begin to become long enough for them to acquire significant fractions of the momentum they possess in the limit where $V \gg V_c$ and viscosity acquires its minimum value η_0 consistent with the temperature. At lesser values of V, since mean free paths are determined by cross sections, we write for the contribution of random thermal molecular momentum to the viscosity of a gas,

$$\eta_b = [1 - (0.5V_c/V)^{2/3}]\eta_0. \tag{3}$$

η_a and η_b overlap beyond $0.5V_c$ while η_a approaches 0 and η_b approaches η_0.

A value of η_0 for a substance at a particular temperature can be determined from a single accurate experimental value of its viscosity as gas at a volume V by calculating η_a for that value of V, subtracting it from η_{obs} and dividing the difference by the quantity in brackets in eq. (3).

We illustrate the applicability of this analysis with excellent data for propane collected and supplemented by Starling, Eakin, and Ellington[5]), and for carbon dioxide by Michels, Botzen and Schuurman[6]).

TABLE I

Parameters for the viscosity of C_3H_8 and CO_2.
Volumes in cm^3/mol, B in reciprocal centipoise and η in centipoise

	V_0	V_c	B	V_0/V_c	η_0 at T	
C_3H_8	61.0	203.2	22.0	0.300	0.0118	411 K
CO_2	28.4	94.0	15.35	0.302	0.0184	348 K

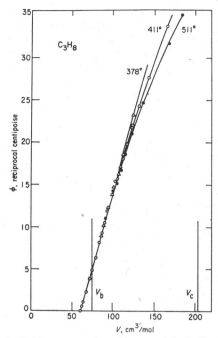

Fig. 1. Fluidity *versus* volume for propane in the liquid region.

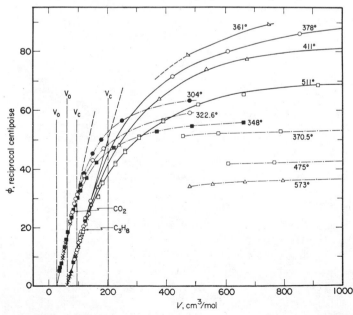

Fig. 2. The fluidity of propane and carbon dioxide along continuous paths from liquid to gas.

The parameters that we have determined are shown in table I. They differ considerably for these two fluids and therefore put appropriate strain upon our concept of the problem. Tables II and III present, for C_3H_8 and CO_2, respectively,

TABLE II

Values of η_a, η_b, $\eta_{calc.}$ and $\eta_{obs.}$ centipoise, at a series of molal volumes (cm^3/mol) for propane at 411 K

V	88.1	115.5	123.9	131.4	164.9	236.6	373.4
η_a	0.1025	0.0508	0.0441	0.0396	0.0267	0.0158	0.0089
η_b	0	0.0009	0.0015	0.0019	0.0033	0.0051	0.0069
$\eta_{calc.}$	0.1025	0.0517	0.0456	0.0415	0.0300	0.0209	0.0158
$\eta_{obs.}$	0.1050	0.0518	0.0453	0.0412	0.0312	0.0208	0.0157
V	534.5	769.1	1022.6	1440	2272	7209	33524
η_a	0.0052	0.0039	0.0029	0.0021	0.0013	0.0004	0.0001
η_b	0.0079	0.0087	0.0092	0.0098	0.0103	0.0111	0.0111
$\eta_{calc.}$	0.0131	0.0126	0.0121	0.0119	0.0116	0.0115	0.0112
$\eta_{obs.}$	0.0135	0.0127	0.0123	0.0119	0.0116	0.0114	0.0110

values of η_a and η_b [calculated with these parameters by means of eqs. (2) and (3)] at those values of molal volume at which "observed" values of viscosity have been reported, shown in the bottom row. The degree of agreement between $\eta_{calc.} = \eta_a + \eta_b$ and $\eta_{obs.}$ is seen by comparing the last two rows.

TABLE III

Values for carbon dioxide corresponding to those in table II. $T = 348$ K

V	52.72	63.96	97.08	107.5	128.2	163.6	220.6	367.6	489.9
η_a	0.0762	0.0522	0.0270	0.0234	0.0186	0.0137	0.0097	0.0054	0.0040
η_b	0.0014	0.0034	0.0071	0.0078	0.0090	0.0104	0.0117	0.0137	0.0142
$\eta_{calc.}$	0.078	0.0556	0.0341	0.0312	0.0276	0.0241	0.0214	0.0191	0.0182
$\eta_{obs.}$	0.076	0.0556	0.0339	0.0311	0.0274	0.0240	0.0214	0.0191	0.0183

Figs. 3 and 4 show the course of η_a and η_b and their sum, $\eta_{calc.}$, the line on which the observed points lie. Deviations of the observed points from this line are, in general, too small to show upon this scale. The figures do not include points for values at either end of the range; the end below V_c approaches asymptotically to the line for V_0; at high values of V, η_a approaches 0, and η_b approaches η_0.

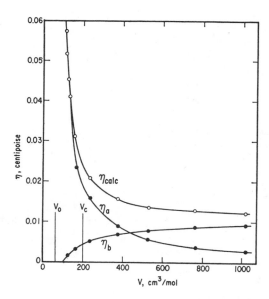

Fig. 3. Viscosity of propane at a series of molal volumes: experimental points ○; contribution of molal crowding, $V - V_0$, and of molecular absorption of momentum, η_a; contribution from thermal momenta at T, η_b; and their sum, the line on which the observed points lie. The solid points represent values calculated at the same volumes as the observed values.

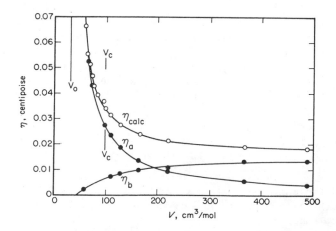

Fig. 4. Viscosity of carbon dioxide at a series of molal volumes. Like fig. 3 but with different scales.

If no experimental value of viscosity of a gas at a particular volume is a available, it is possible to calculate η_0 from V_0 as follows.

Kinetic theory has yielded eq. (4) for the viscosity of a hard-sphere gas at low density.

$$\eta_0 = 2\,(mkT)^{\frac{1}{2}}/3\pi^{3/2}\sigma^2, \tag{4}$$

where m is molecular mass, k is the Boltzmann constant, and σ is molecular diameter. Replacing the imprecise σ^2 by $V_0^{2/3}$, and collecting invariant terms and conversion factors into a constant, K, permits writing eq. (5), in molal units,

$$\eta_0 = K(MT)^{\frac{1}{2}}/V_0^{2/3}. \tag{5}$$

Its validity can be tested with the data for C_3H_8 and CO_2 given in table I. (Molal weights are the same.)

Distinguishing the quantities for CO_2 by primes, we write

$$\eta_0/\eta_0' = (T/T')^{\frac{1}{2}}\,(V_0'/V_0)^{2/3}.$$

The left-hand member yields the value 0.642, that on the right gives 0.653. The near agreement is striking, in view of the differences between the parameters and temperatures, and the different laboratories in which these two gases were investigated.

It seems probable, therefore, that eq. (5) will be found to be generally valid for simple species of fluids. The data for C_3H_8 give $K = 1.36 \times 10^{-3}$; those for CO_2 give 1.38×10^{-3}. Even this small discrepancy may be related to the fact that these two fluids diverge unequally from the hard-sphere model, as shown by their values of $1/B$, 0.045 for C_3H_8, 0.074 for CO_2.

This treatment removes η_0 as a separate parameter and makes viscosity over the whole range, from V_0 to dilute gas, calculable from the parameters V_0 and B.

We offer the foregoing as a contribution to knowledge about the liquid state and also as a practical method for dealing with the hitherto nearly intractable region of compressed gases. It is not the final word on their viscosity, but it is a treatment that can be used and further refined, since it involves no assumption that is inconsistent with what is known about simple fluids.

Acknowledgement. Acknowledgement is made to donors of the Petroleum Research Fund, administered by the American Chemical Society, for its support of this research.

REFERENCES

1) Hildebrand, J.H., Science **174** (1971) 490.
2) Batschinski, A.J., Z. physik Chem. **84** (1913) 643.
3) Hildebrand, J.H. and Lamoreaux, R.H., Proc. Nat. Acad. Sci. **69** (1972) 3428.
4) Hildebrand, J.H. and Lamoreaux, R.H., J. phys. Chem. **77** (1973) 1471.
5) Starling, K.E., Eakin, B.E. and Ellington, R.T., A.L. Ch. E. Journal **6** (1960) 438.
6) Michels, A., Botzen, A. and Schuurman, W., Physica **23** (1957) 95.

7

Solubility of Gases in Liquids: Fact and Theory

J. H. Hildebrand* and R. H. Lamoreaux

Department of Chemistry, University of California, Berkeley, Calif. 94720

Values of entropy involved in transferring one mole of different nonreacting gases at 25°C and 1 atm into a nonpolar liquid saturated at mole fractions x_2 when plotted as $\bar{S}_2 - \bar{S}_2^g$ against values of $-R \ln x_2$, give a straight line intercepting the entropy axis at -21.0 cal deg^{-1} mol^{-1}. All solvents give lines with the same intercept, including $(C_4F_9)_3N$ with fluorocarbon gases, but not with alkanes or the rare gases. Defining excess partial molal entropy as $\Delta \bar{S}_2^E = \bar{S}_2 - \bar{S}_2^g + 21.0$, the slope of the line for a particular solvent, $\Delta \bar{S}_2^E / -R \ln x_2$, is a parameter for that solvent. It is a function of its cohesion as measured by its energy of vaporization per cubic centimeter, $\Delta E_1^v / V_1$, which is the square of its solubility parameter δ_1. We consider $\Delta \bar{S}_2^E$ as the sum of *dilution entropy*, $-R \ln x_2$, and *configuration entropy*, $\Delta \bar{S}^c$. This component comes from the weak attractive potential between a gas molecule and its surrounding solvent molecules which causes expansion, often very large, and enhanced freedom of thermal motion. An approximately inverse relation between expansion and solvent cohesion makes $\Delta \bar{S}_2^c$ nearly uniform, with certain exceptions, for each gas in different solvents. The distribution of points for $T\Delta \bar{S}^c$ for the various gases is determined by the difference between $T(21.0 - R) = 5.67$ kcal mol^{-1} and the energy of vaporization of each gas at its boiling point. The resulting points fall on a single straight line. This line permits calculation of gas solubilities from only two parameters, $\Delta E_1^v / V_1$ for solvents and ΔE_2^v for gas solutes. We discuss departures from these relations and point out that a fully satisfactory treatment of the subject awaits a refinement of the theory of London for dispersion forces to yield a valid relation for forces between like and unlike molecules.

Introduction

This paper presents a theoretical analysis of the body of information now at hand about solutions of nonreacting gases in nonpolar liquids. A general theory valid for gases whose properties span the enormous range from helium to c-C_4F_8 cannot be constructed in disregard of any of the pertinent properties, such as liquid structure, entropy of solution, partial molal volumes, molecular structures, and cohesion of the solvent and solute in their pure liquid state. As Alfred North Whitehead has written, "There can be no true physical science which looks first to mathematics for the provision of a physical model."

Reprinted from **I&EC FUNDAMENTALS**, Vol. 13, Page 110, May 1974

We have carried on many investigations designed to reveal facts that a satisfactory theory should be able to correlate and explain. These are listed in Appendix I and followed in Appendix II by a list of other important sources.

Solubility and Entropy of Solution

Values of the entropy of solution for gases that obey Henry's law are obtained from the change of solubility with temperature by the equation (Hildebrand, *et al.*, 1971)

$$\overline{S}_{|2} - S_2{}^g = R\left(\frac{\partial \ln x_2}{\partial \ln T}\right)_{P,\,\text{sat.}} \tag{1}$$

Plots of log x_2 against log T for gases at 1 atm in nonpolar liquids well below their boiling points yield straight lines (Hildebrand, *et al.*, 1971) whose slopes multiplied by the gas constant give values of the partial molal entropy of solution. These values, plotted against values of solubility at 25°C and 1 atm, expressed as $-R \ln x_2$, the dilution, which increases with unlikeness of attractive potential of solvent and solute, give the grid of lines represented in Figure 1. It includes, for illustration, only liquids in which we have determined solubility and entropy of solution of many gases with satisfactory accuracy. These liquids were chosen to cover a wide range in solvent power. Their "cohesive energy density," (Scatchard) energy of vaporization per cubic centimeter, varies from 35 cal cm^{-3} for $(C_4F_9)_3N$ to 100 cal cm^{-3} for CS_2.

A Common Reference State for Partial Molal Entropy of Solution of Gases

The lines for the solvents in Figure 1 all converge to -21.0 cal deg^{-1} mol^{-1} at $-R \ln x_2 = 0$. This value has the following significance. The succession of points on descending the line for any solvent represent gases whose attractive potentials approach progressively more and more closely to that of the solvent. Also, their mole fractions x_2 increase and x_1 for the solvent decrease until, at the intercept $x_2 = 1$ and $x_1 = 0$, and the value -21 cal deg^{-1} mol^{-1} represents the entropy of condensing a gas whose boiling point is 25°C at 1 atm. The "Hildebrand Rule" (Hildebrand, *et al.*, 1971) permits calculation of the entropy of vaporization of such a liquid. Hermsen and Prausnitz (1961), using this rule, calculated the entropy of vaporization of 21 nonpolar liquids at a temperature where their corrected vapor volumes are 49.5 l. mol^{-1}, getting a mean value of 22.3 $\pm$ 0.08 cal deg^{-1} mol^{-1}. Changing the vapor volume to 24.5 1./mol gives ΔS^v exactly 21.0 cal deg^{-1} mol^{-1} and $\Delta E^v = 5.67$ kcal mol^{-1}.

It is significant that the points for fluorocarbon gases fall on the line for $(C_4F_9)_3N$ but points for other gases do not. Although these gases become more soluble in order of their boiling points, none are sufficiently "like" the solvent for their points to stay on the line. These are some of many pieces of evidence that intermolecular forces can differ with orbital types as well as strengths.

It is obviously appropriate to use the intercept at 21.0 cal deg^{-1} mol^{-1} as a base for the partial molal entropy of a gas; accordingly, we will call the excess entropy above -21.0 the *excess partial molal entropy*

$$\Delta \bar{S}_2^{E} = (\bar{S}_2 - S_2^{g}) + 21.0 \qquad (2)$$

One must remember that 21.0 represents the intercept for solubility measured at 25°C and 1 atm; therefore the reference substance is one whose boiling point is 298.2 K, where its vapor volume is 24.5 l. At a lower temperature, e.g., 283.2 K, the vapor molal volume would be 25.8 l. and its entropy of vaporization would be 21.0 + R ln (25.8/24.5) = 21.56. The intercept would be at -21.56 cal deg^{-1} mol^{-1} and all the lines would be rotated toward the vertical about the points where they cross $\bar{S}_2 - S_2^{g} = 0$.

It has been our custom to distinguish properties of solvent and solute by subscripts 1 and 2, but in what follows the latter can be omitted to simplify typography.

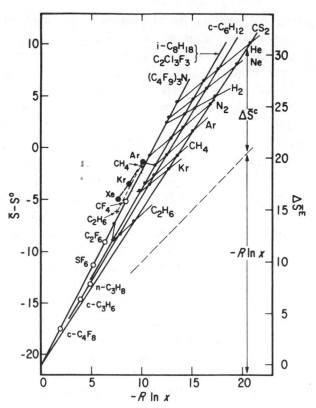

Figure 1. Entropy of solution of gases at 25° and 1 atm into representative solvents plotted against their solubility expressed as $-R \ln x$ (unit, cal deg^{-1} mol^{-1})

Entropy of Dilution and Configuration

If the difference in entropy of solution of two gases were only the result of different values of mole fraction, then all points would fall upon the dashed line of unit slope in Figure 1. We see, however, that $\Delta \bar{S}^{E}$ is larger than $-R \ln x$ by an amount that we shall call *configuration entropy*, $\Delta \bar{S}^{c}$; hence

$$\Delta \bar{S}^{E} = -R \ln x + \Delta \bar{S}^{c} \qquad (3)$$

$\Delta \bar{S}^{c}$, the nondilution component of $\Delta \bar{S}^{E}$, results from the fact that solvent molecules surrounding a solute molecule of small attractive potential are held less tightly and gain freedom of thermal motion. The accompanying expansion is seen in partial molal volumes far exceeding molal volumes of the solute gases at their boiling points. In the case of CF_4 in c-C_6H_{12}, for example, $\bar{V}/V_b = 1.54$. For a given gas, this ratio is larger the smaller the cohesion of the solvent; therefore slopes, S, of solvent lines are related to solvent parameters as seen in Table I and Figure 2. The empirical equation is

$$S = 0.90 - 4.0\delta_1{}^2 \tag{4}$$

A major discrepancy exists in the case of $(C_4F_9)_3N$ and other perfluorochemical solvents. Points for $2,2,4$-C_8H_{18} (i-C_8H_{18}) and $CCl_2F \cdot CClF_2$ fall on the same line although their solubility parameters differ.

Solubility of One Gas in Different Solvents

The lines for solvents in Figure 1 are so crowded that we have plotted in Figure 3 $-RT \ln x_2$ against $\delta_1{}^2$, using for the latter a scale five times the scale for the former. We see a grid of relations upon which most gases fit so well as to make it serve with considerable confidence both to predict missing values and to correct those which appear to be inaccurate, especially Xe in several liquids, He in

Table I. Slopes of the Lines in Figure 1 and Values of $\Delta E_1/V_1 = \delta_1{}^2$ for the Several Solvents

	S	$\delta_1{}^2$ cal/cm^3
$(C_4F_9)_3N$	1.88	35
$2,2,4$-$(CH_3)_3C_5H_9$	1.70	47
$CCl_2F \cdot CClF_2$	1.70	52
n-C_7H_{16}	1.68	55
c-$C_6H_{11}CH_3$	1.65	62
c-C_6H_{12}	1.63	67
CCl_4	1.60	74
C_6H_6	1.56	84
CS_2	1.51	100

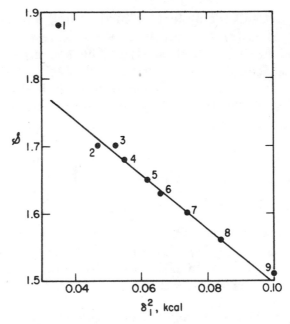

Figure 2. Relation between the slopes of lines such as those in Figure 1 for solubility in representative solvents and their solubility parameters. Sequence: (1) $(C_4F_9)_3N$; (2) $CCl_2F\cdot CClF_2$; (3) i-C_8H_{18}; (4) C_7H_{16}; (5) c-$C_6H_{11}CH_3$; (6) c-C_6H_{12}; (7) CCl_4; (8) C_6H_6; (9) CS_2

CCl_4, and Kr in CS_2. We see again the divergence of $(C_4F_9)_3N$ from other solvents.

Values of Configuration Entropy

In Table II are given values of $T\Delta S^c$ and $-RT \ln x$ for most of the common gases and solvents; ΔS^c and $-R \ln x$ have been multiplied by $T = 298°C$ in order to convert to the dimension of energy for later correlation with the parameters we use for gases.

The striking feature of this table is that values of ΔS^c for a single gas in different solvents, except H_2 and CH_4, vary from solvent to solvent little more than the uncertainties of experimental data. This corresponds to the near parallelism between the lines for the gases and the dotted line for $-R \ln x$ in Figure 1.

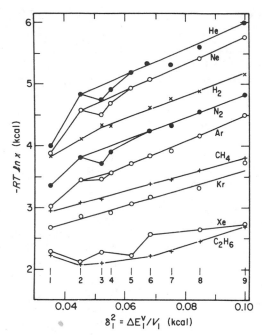

Figure 3. Solubility of a gas, expressed as $-RT \ln x$ in a series of solvents plotted against values of δ_1^2. Sequence: same as in Figure 2

One may wonder why values of $T\Delta\bar{S}^c$ are not significantly different in solvents which differ considerably in cohesion. We offer the following explanation.

The physical interpretation of configuration entropy is that the attractive potentials of the gas molecules are much less than those of the solvent, and this permits expansion and greater freedom of thermal motion of the surrounding molecules of solvent. This expansion can be very large. Because we have no precise measure of intrinsic molecular volumes, we use the molal volumes of gases at their boiling points, V_b (Hildebrand, 1969; Hildebrand, *et al.*, 1971, p 217). These are uniform fractions of critical volumes and are free from the experimental uncertainty of values of σ derived from PVT data. The answer is to be found in an equation, whose derivation is given by Hildebrand, *et al.* (1971, pp 79, 80, 184–187), for the excess of

Table II. Values of $-RT \ln x$ (First Row) and $T\Delta\bar{S}^c$ (Second Row)

ΔE^v			$i\text{-}C_8H_{18}$	$C_2Cl_3F_3$	$n\text{-}C_7H_{16}$
0.175	14	H_2	4.22	4.33	4.31
			2.96	3.03	2.97
0.018	13	He	4.84	4.74	4.91
			3.38	3.31	3.39
0.37	12	Ne	4.56	4.50	4.70
			3.19	3.16	3.25
1.18	11	N_2	3.83	3.71	3.90
			2.68	2.60	2.69
1.28	10	CO			
1.385	9	Ar	3.45	3.46	3.55
			2.41	2.42	2.45
1.45	8	O_2	3.48		
			2.43		
1.73	7	CH_4	3.08	3.13	
			2.15	2.19	
2.00	6	Kr	2.86		2.91
			2.00		2.01
2.69	5	Xe	2.12	2.28	
			1.48	1.60	
2.90	4	C_2H_4			
3.15	3	C_2H_6	2.09	2.10	
			1.46	1.47	
4.03	2	C_3H_8			
4.31	1	$c\text{-}C_3H_6$			

the partial molal volume of a highly dilute solute over its molal volume in the pure liquid state. The equation is

$$(\bar{V} - V_0)(\partial S_1/\partial V)_T = \bar{S} - S_0$$

or the equivalent form

$$(\bar{V} - V_0)(\partial E_1 \partial V)_T = \Delta\bar{E}_2$$

To apply it to solutions of gas we have used for V_0 the molal volume of the gas at its own boiling points—a "corresponding states" value (Hildebrand, 1969; Hildebrand,

c-C$_6$H$_{11}$CH$_3$	c-C$_6$H$_{12}$	CCl$_4$	C$_6$H$_6$	CS$_2$
	4.61	4.77	4.89	5.15
	2.90	2.86	2.73	2.63
5.16	5.33	5.34	5.60	6.00
3.35	3.36	3.21	3.14	3.06
4.94	5.08		5.40	5.76
3.21	3.20		3.02	2.94
	4.24	4.33	4.55	4.83
	2.67	2.60	2.55	2.46
		4.18	4.33	4.69
		2.51	2.42	2.40
3.72	3.84	3.92	4.15	4.52
2.42	2.42	2.35	2.39	2.31
		3.86	4.21	4.56
		2.32	2.36	2.33
	3.38	2.46	3.61	3.80
	2.13	2.08	2.02	1.94
3.05	3.17		3.31	3.74
1.98	2.00		1.86	1.91
2.23	2.57		2.66	2.71
1.45	1.62		1.41	1.40
		2.51	2.38	
		1.50	1.46	
	2.21	2.28	2.48	2.68
	1.39	1.41	1.39	1.37
	1.43		1.69	
	0.90		0.95	
	1.17		1.25	
	0.74		0.71	

et al., 1971, p 217).

Table III gives values of $\bar{V} - V_b$ for Ar in a series of solvents of decreasing internal pressure, $(\partial E_1/\partial V)_T$, using the values given by Hildebrand, *et al.* (1971, p 217). V_b for Ar is 28.6 cm^3. Values of $\bar{V}$ were measured by Hildebrand and Jolley (J), and by Hildebrand and Walkley (W). The mean value of the product is less than $T\Delta\bar{S}^c$ for argon (see Table II), $298 \times 8.1 = 2.40$ kcal mol^{-1}, which seems to indicate that the configuration entropy involves more than

Table III. Reciprocal Relation between Expansion, $\bar{V} - V_b$ (cm³ at 25°) and Internal Pressure of Solvent for Ar ($V_b = 28.6$ cm³)

	$\bar{V}$		$\bar{V} - V_b$	$(\partial E_1/ \partial V)_T$	Product
C_6H_6	44.6	W	16.0	88.4	1410
CCl_4	44	J	15.4	81.1	1250
c-C_6H_{12}	47.6	J	19.0	75.5	1430
n-C_7H_{16}	48.3	W	19.7	60.7	1200
c-$C_6F_{11}CF_3$	51	J	22.4	54.4	1220

Table IV. Partial Molal Volumes, V (cm³/mol) of Fluorochemical Gases in c-C_6H_{12}; Boiling Point Volumes, V_b (cm³), Relative Expansions, $(V - V_b)/V_b$, and Solubility as $-R \ln x$, All at 25°

	c-C_4F_8	C_3F_8	SF_6	C_2F_6	CF_4
$\bar{V}$ (measd)	144	140	101	111	86
V_b	123.4	117.0	75.5	86.0	54.3
$(\bar{V} - V_b)/V_b$	0.17	0.20	0.30	0.29	0.58
$-R \ln x$	7.62	10.13	10.35	11.93	13.55
$\bar{V}$ (calcd)	142	142	101	116	88

just expansion. The values in the last column of the product of expansion by internal pressure are approximately constant; expansion increases as cohesion of the solvent decreases.

Partial molal volumes of fluorochemical gases in cyclohexane, shown in Table IV, were measured by Linford and Hildebrand (1969). Miller and Hildebrand (1968) had determined their solubility. Relative expansions, $(\bar{V} - V_b)/V_b$ range from 17 to 58%. Values of $\bar{V}$ calculated by eq 4 with $\mathcal{S} - 1 = 0.63$ and $(\partial S_1/\partial V)_T = 0.253$ cal deg cm³ mol are in the bottom row of the table. They agree closely with the measured values in the top row.

Prediction of the Solubility of Individual Gases

The foregoing sections have dealt principally with relative solvent powers of different solvents. We saw that the

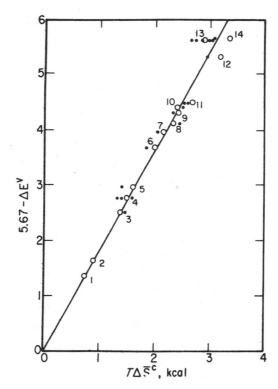

Figure 4. Ordinates, partial molal energy of solution calculated as $5.67 - \Delta E^v$; abscissas, conformation entropy, $T\Delta S^c$

ratio of configuration entropy to dilution entropy is virtually the same, with certain exceptions, for all gases in the same solvent. There remains the problem of predicting the positions of the points for individual gases along the line for a given solvent.

We mentioned earlier that the successive points in descending a solvent line in Figure 1 are for gases that are successively more like the solvent until, at $-R \ln x = 0$, solute and solvent become one, with entropy of vaporization 21.0 cal deg^{-1} mol^{-1} and energy of vaporization 5.67 kcal mol^{-1}. The most practical parameters for the solubility of gases are their energies of vaporization at their boiling points (Hildebrand, 1969). Accordingly, we adopt

$5.67 - \Delta E_2^v$ as the ordinate of the plot of the several gases in Figure 4. Values of $\Delta \bar{E}_2^v$ are given in Table II. As abscissas, we use values of $T \Delta \bar{S}^c$, which, as we have seen, make the points for a gas in different solvents virtually coincide, except in the case of a fluorochemical solvent. Points for many solutions are plotted with these coordinates in Figure 4. The open circles represent values in cyclohexane, which we consider to be the most accurate. The solid points are for the same gas in other solvents; their spread we think is the result of experimental errors, as is surely the case with Xe, but not in the case of H_2, point 15, where the drift seen in Table II is certainly real, and a challenge to theory. The slope of the line is 1.81. It is evident that this relation permits the prediction of the solubility of a gas from its energy of vaporization at its boiling point. Its ordinate on a line of slope 1.81 gives a value for $T \Delta \bar{S}^c$, which, divided by $\mathcal{S} - 1$ for a particular solvent, gives the value of $-R \ln x$. From the location of this value upon the corresponding line in Figure 1, one can obtain the change of solubility with temperature. Those who regard geometry as aesthetically inferior to algebra can readily transform that process into a single equation if they wish, but we have preferred a graphic exposition in order to make clearer the physical steps involved. It should be evident that it is possible, with two nonadjustable parameters, $\Delta E_1^v / V_1$ and ΔE_2^v, to calculate with fair approximation the solubility of any inert gas except H_2 in any nonpolar solvents except fluorochemicals.

SF$_6$ and CF$_4$

These two gases are not included in Figure 1 because they depart widely from the grid of relations there shown. In Figure 5 are plotted their solubility as $-RT \ln x$, scale on left axis, in the usual list of solvents against their values of δ_1^2. The data are shown in Table V. On this kind of plot most inert gases give sloping straight lines, as

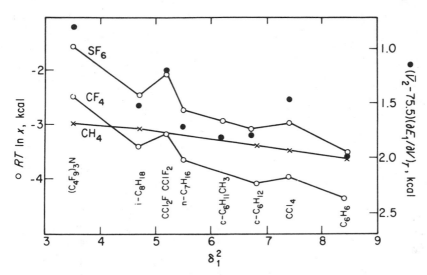

Figure 5. Illustration of divergent solubility of SF_6 and CF_4 in halogen containing alkane solvents, and inverse relation of solubility and partial molal volume of SF_6

Table V. Solubility SF_6 as $-RT \ln x$ (kcal) and Its Partial Molal Volume at 25°C, $\bar{V}$ (cm³), Internal Pressure of the Solvents, $(\partial E/\partial V)_T$ (cal/cm³), and Calculated Values of Partial Molal Energy of Solution, $\Delta\bar{E}$(kcal)

	$-RT$ $\ln x$	$\bar{V}$	$(\partial E_1/ \partial V)_T$	$\Delta\bar{E}$, kcal
$(C_4F_9)_3N$	1.55	91.3	52	0.82
$C_2Cl_3F_3$	2.09	94.8	62.1	1.20
$i\text{-}C_8H_{13}$	2.48	102.7[a]	55.1	1.60
$n\text{-}C_7H_{16}$	2.72	103.2	60.7	1.68
$c\text{-}C_6H_{11}CH_3$	2.93	101.6	69.5	1.82
$c\text{-}C_6H_{12}$	3.09	101.4[a]	70.0	1.81
CCl_4	2.98	94.7	81.0	1.55
C_6H_6	3.52	97.9[b]	88.4	1.98

[a] Walkley and Jenkins (1968). [b] Walkley and Jenkins (1968) obtained 97.1.

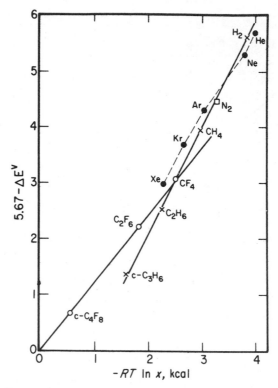

Figure 6. Conformity of fluorocarbon gases and divergence of rare gases and alkanes to solubilities in $(C_4F_9)_3N$

seen in Figure 2 and as illustrated here by the line for CH_4, whereas SF_6 and CF_4 give lines that zig-zag in parallel, with peaks of solubility at $(C_4F_9)_3N$, $C_2Cl_3F_3$, and CCl_4.

The solid points are for partial molal volumes of SF_6 in excess of 75.5 cm³, the extrapolated value of its boiling point volume. (It is solid at 1 atm.) The scale on the right-hand vertical axis increases downward in order to illustrate that expansion is larger in the poorer solvents, and *vice versa*.

$(C_4F_9)_3N$ as Solvent for Gases

Because of the divergence of the solubility of all but

fluorochemical gases in a fluorochemical solvent, such systems are not included in the foregoing treatment, but are represented in Figure 6 in a plot of $5.67 - \Delta E^v$ against $-RT \ln x$. One sees that CF_4, C_2F_6, and c-C_4F_8 fall on a straight line whose origin is at $5.67 - \Delta E^v$ zero, as in Figure 4 and whose slope is 1.21. The rare gases and the alkanes do not fall upon the line, indicating that attractive potentials between a fluorochemical solvent and the gases of these two groups diverge from the pattern common to the components of the solutions represented in Figure 1. The lack of an adequate theory for these irregular interactions is the major block on the road to progress toward a general treatment of solutions of nonpolar, nonreactive solutions of gases, liquids, and solids in liquids.

Critique

We wish to emphasize several features of the foregoing survey.

1. The dissolved gases in all the solutions we have considered are so dilute that interactions between solute molecules can be neglected, as shown by their conformity to Henry's law. For this reason claims that one chemically inert gas can change the solubility of a second are incredible.

2. It is sufficient to regard a dissolved molecule of gas as being in a field of force instead of in a sum of pair potentials with individual solvent molecules at distances that are simple functions of the diameters, σ, of the molecules of gas. It is not practical to assume distribution functions that can serve to calculate the energy and entropy of individual molecules. It seems absurd to derive formulas whose parameters are not actually operational. Partial molal volumes of the less soluble gases far exceed their liquid molal volumes at their boiling points.

Acknowledgment. Acknowledgment is made to the donors of The Petroleum Research Fund, administered by

the American Chemical Society, for support of this research.

Appendix I. Publications by J. H. Hildebrand and Collaborators Dealing with Solutions of Inert Gases in Liquids

1. Solubility relations of certain gases, with N. W. Taylor, *J. Amer. Chem. Soc.*, **45**, 682 (1923).

2. Book: "Solubility," Chemical Catalog Co., New
York, N. Y., 1924.

3. Possibilities in the use of helium–oxygen mixtures as
a mitigation of caisson disease, with R. R. Sayers and W.
P. Yount, Report of Investigations, U. S. Bureau Mines,
Series No. 2670, Feb 1925.

4. Solubility of N_2 in CS_2, C_6H_6, n-C_6H_{14}, c-C_6H_{12},
and 3 fluorocarbons, with J. Chr. Gjaldbaek, *J. Amer.
Chem. Soc.*, **71**, 3147 (1949).

5. Solubility of Cl_2 in C_7F_{16} and other liquids, with J.
Chr. Gjaldbaek, *J. Amer. Chem. Soc.*, **72**, 609 (1950).

6. Some partial molal volumes of gases in solution, with
J. Chr. Gjaldbaek, *J. Amer. Chem. Soc.*, **72**, 1077 (1950).

7. Solubility and entropy of solution of Ar in five selected nonpolar solvents, with L. W. Reeves, *J. Amer. Chem.
Soc.*, **79**, 1313 (1957).

8. Solubility, entropy, and partial molal volumes in solutions of gases in nonpolar solvents, with J. E. Jolley, *J.
Amer. Chem. Soc.*, **80**, 1050 (1958).

9. Partial molal volumes of hydrogen and deuterium,
with J. Walkley, *J. Amer. Chem. Soc.*, **81**, 4459 (1959).

10. Solubility and entropy of solution of He, N_2, Ar, O_2,
CH_4, C_2H_6, CO_2, and SF_6 in various solvents, regularity
of gas solubilities, with Y. Kobatake, *J. Phys. Chem.*, **65**,
331 (1961).

11. Solubility of CF_4 in $C_6F_{11}CF_3$, with L. W. Reeves,
J. Phys. Chem., **67**, 1918 (1963).

12. Solubility and entropy of solution of CF_4 and SF_6 in
nonpolar solvents, with G. Archer, *J. Phys. Chem.*, **67**,
1830 (1963).

13. Partial molal volumes of SF_6, with H. Hiraoka, *J. Phys. Chem.*, **67,** 1919 (1963).

14. Solubility and entropy of solution of certain gases in $(C_4F_9)_3N$, $CCl_2F \cdot CClF_2$ and $2,2,4-(CH_3)_3C_5H_9$, with H. Hiraoka, *J. Phys. Chem.*, **68,** 213 (1964).

15. Apparatus for accurate, rapid determination of the solubility of gases in liquids, with J. H. Dymond, *Ind. Eng. Chem., Fundam.*, **6,** 130 (1967).

16. Solubility of a series of gases in $c-C_6H_{12}$ and $(CH_3)_2SO$, with J. H. Dymond, *J. Phys. Chem.*, **71,** 1829 (1967).

17. Solutions of gases in normal liquids, J. H. Hildebrand, *Proc. Nat. Acad. Sci. U. S.*, **57,** 549 (1967).

18. Solutions of inert gases in water, with K. W. Miller, *J. Amer. Chem. Soc.*, **90,** 3001 (1968).

19. Solubility of fluorocarbon gases in $c-C_6H_{12}$, with K. W. Miller, *J. Phys. Chem.*, **72,** 2248 (1968).

20. Partial molar volumes of fluorochemical gases in $c-C_6H_{12}$, with R. G. Linford, *Trans. Faraday Soc.*, **65,** 1470 (1969).

21. Solubility of gases in mixtures of nonpolar liquids, with R. G. Linford, *J. Phys. Chem.*, **73,** 4410 (1969).

22. Solubility and entropy of solution of gases in $CCl_2F \cdot CClF_2$, with R. G. Linford, *Trans. Faraday Soc.*, **66,** 577 (1970).

23. Thermodynamic parameters for dissolved gases, J. H. Hildebrand, *Proc. Nat. Acad. Sci. U. S.*, **64,** 1331 (1969).

24. Solubility and entropy of solution of 16 gases in $(C_4F_9)_3N$ and CS_2, R. J. Powell, *Chem. Eng. Data*, **17,** 302 (1972).

Appendix II. Papers by Other Important Sources

1. J. Horiuti, *Sci. Papers, Inst. Chem Res., Tokyo*, **17,** No. 341, 125 (1931).

2. H. W. Cook, D. N. Hansen, and B. J. Alder, *J. Chem. Phys.*, **26,** 748 (1957).

3. A. Lannung, *J. Amer. Chem. Soc.*, **52,** 68 (1930).

4. H. L. Clever, R. Battino, J. H. Saylor, and P. N. Gross, *J. Phys. Chem.*, **61,** 1078 (1957); **62,** 89, 375, 1334 (1958).

5. J. C. Gjaldbaek, *Acta Chem. Scand.*, **6,** 623 (1952); **8,** 1398 (1954).

6. J. C. Gjaldbaek and H. Nieman, *Acta Chem. Scand.*, **12,** 611 (1958).

7. A. Lannung and J. C. Gjaldbaek, *Acta Chem. Scand.*, **14,** 1124 (1960).

8. E. S. Thomsen and J. C. Gjaldbaek, *Acta Chem. Scand.*, **17,** 127 (1963).

9. R. Battino and H. L. Clever, *Chem. Rev.*, **66,** 395 (1966).

10. E. Wilhelm and R. Battino, *Chem. Rev.*, **73,** 1 (1973).

Literature Cited

Hermsen, W., J. M., Prausnitz, *J. Chem. Phys.*, **34,** 108 (1961).

Hildebrand, J. H., *Proc. Nat. Acad. Sci. U. S.*, **64,** 1331 (1969).

Hildebrand, J. H., Prausnitz, J. M., Scott, R. L., "Regular and Related Solutions," p 40, Van Nostrand-Reinhold, New York, N. Y., 1971.

Linford, R. G., Hildebrand, J. H., *Trans. Faraday Soc.*, **65,** 1470 (1969).

Miller, K. W., Hildebrand, J. H., *J. Phys. Chem.*, **72,** 2248 (1968).

Walkley, J., Jenkins, W. I., *Trans. Faraday Soc.*, **64,** 19 (1968).

Received for review July 25, 1973
Accepted January 14, 1974

8

Diffusivity of Gases in Liquids

(carbon tetrachloride/perfluorotributyl amine/temperature)

J. H. HILDEBRAND AND R. H. LAMOREAUX

College of Chemistry, University of California, Berkeley, Calif. 94720

Contributed by J. H. Hildebrand, May 30, 1974

ABSTRACT Diffusion coefficients of H_2, Ne, N_2, Ar, CH_4, Cl_2, CF_4, C_2H_6, SF_6, I_2, and isotopic CCl_4, all in CCl_4, determined at atmospheric pressure, are linear functions of temperature, converging to zero at the temperature where CCl_4 ceases to be fluid. The slopes of these lines increase with decreasing molecular cross-section of the diffusants, and with increasing entropy of expansion of the diffusants in CCl_4. Diffusivities in $(C_4F_9)_3N$, whose molecules are very large and three-armed, do not converge as temperature is decreased. Molecules of H_2, Ne, and, to a lesser extent, Ar, are able to diffuse in $(C_4F_9)_3N$ even at temperatures where fluidity is low.

The work here reported is a major extension of a series of investigations of diffusion carried on in this laboratory. They are: diffusion of iodine in carbon tetrachloride under pressure (1); self-diffusion of carbon tetrachloride (2); diffusion of H_2, D_2, N_2, CCl^*_4 (isotopic), Ar, CH_4, and CF_4 in CCl_4 (3); quantum effect in the diffusion of gases in liquids at 25° (4); and

Reprinted from
Proc. Nat. Acad. Sci. USA
Vol. 71, No. 9, pp. 3321–3324, September 1974

diffusivity of ^{3}He, ^{4}He, H_2, D_2, Ne, CH_4, Ar, Kr, and CF_4 in $(C_4F_9)_3N$ (5).

In 1971 Hildebrand (6), writing on "motions of molecules in liquids," remarked that since diffusion represents entropy increasing toward the maximum possible under the conditions, it is desirable to study the change of diffusivity with temperature, and further, that diffusivity and fluidity can be expected to begin at the same volume and temperature. We here report, accordingly, the diffusivity and its change with temperature of gases in solution in CCl_4 and in $(C_4F_9)_3N$.

EXPERIMENTAL

The method, described in refs. 3–5, is to measure the volume of gas that passes per second, first into a shallow layer of solvent above a diaphragm, then through the diaphragm, and into a large vessel below filled with the solvent. As soon as the gas has formed a uniform concentration gradient within the diaphragm, a steady state is achieved and the rate of disappearance of the gas, measured by volume at constant pressure and temperature, becomes linear with time.

The diaphragm, described in ref. 4, is a slice, 0.90-cm thick, cut from a cylinder composed of 2962 lengths of stainless steel hypodermic tubing in parallel array and inbedded in solder. The tubes were filled with melted triphenylamine, which, after freezing, permitted a slab to be cut without damage to the ends of the tubes. The volume of these tubes per cm of length was determined with water. The total open area of the diaphragm is 5.68 cm². This area and path length permit us to calculate diffusion coefficients directly in cm²/sec.

The volume of liquid below the diaphragm was 240 cm³. The concentration of the gas in this liquid was negligible. The liquid was gently stirred during an experiment by a magnetic device in order to distribute uniformly the small amount of gas that had passed through the diaphragm.

In order to prevent vapor of the solvent in the glass dome above the diaphragm from diffusing up-stream into the measuring buret, the glass dome and diaphragm are connected by a long glass capillary tube through which the rate of flow of gas exceeds any possible counter-diffusion of solvent vapor. Gas was introduced at a pressure kept constant by balancing against air in a large thermostated flask.

When gas is admitted to the diffusion cell, it first saturates the thin layer of solvent above the diaphragm, then establishes a uniform concentration gradient within the diaphragm, after which $\Delta n/\Delta t$ (moles of gas per sec) is proportional (diffusivity D) to the differences in concentration above and below the diaphragm, ΔC, and to the diffusing area, a, and inversely proportional to the length, l, and

$$D = [l/a\Delta c] \, dn/dt$$

When concentration is expressed as moles per cm^3, D is in units of cm^2/sec.

Values for concentration are calculated from values of mole fraction, and its change with temperature can be found in refs. 10 and 11. For CCl_4, a correction from Henry's law for gas solubility in the solvent layer above the diaphragm must be made to take into account the lowered partial pressure of gas in the diffusion cell. This, and prevention of solvent vapor diffusing into the gas buret, eliminate uncertainties in our earlier work with CCl_4 as solvent. The vapor pressure of $(C_4F_9)_3N$ is less than one torr at 25°.

The CCl_4 was Matheson, Coleman and Bell "Spectroquality." The $(C_4F_9)_3N$ was distilled through a column of many plates; the portion used boiled between 174.7° and 175.5°. The solvents were thoroughly degassed before use by repeatedly pumping on the frozen solvent and melting. The gases were manufacturer's "research grade."

DIFFUSIVITY IN CCl_4

Our results for diffusivity in CCl_4 are given in Table 1 and plotted in Fig. 1. We include data for I_2 from ref. 1, for Cl_2 from ref. 7, and for the self-diffusion of CCl_4 from ref. 8.

One sees that the lines are straight, and converge at $D=0$ to nearly the same temperature, $-31°$. The molal volume of CCl_4 at this temperature is 90 cm³. Its fluidity becomes zero also at 90 cm³/mol (9).

Table 1 also gives values of $V_0^{2/3}$, where V_0 is the intrinsic molal volume obtained by extrapolating to zero fluidity the straight line plots of fluidity against molal volume, except in the cases of CF_4 and SF_6, for which values of fluidity are not

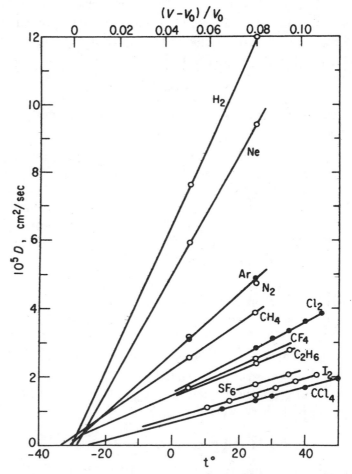

FIG. 1. Variation of diffusivity in centipoise of gases in CCl_4 with temperature and fractional expansion of CCl_4.

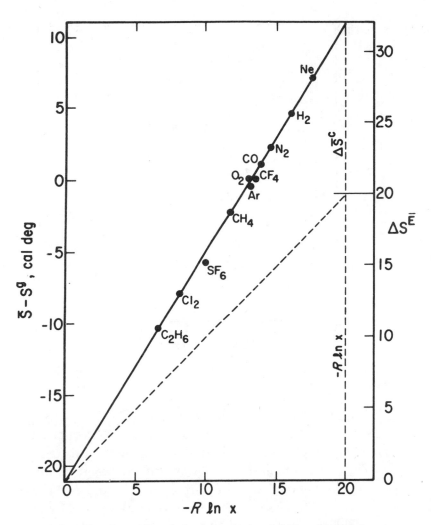

FIG. 2.   Illustration of the separation of excess molal entropy of solution of gases in CCl_4 into molal entropy of dilution and of configuration.

available. The values in the table were calculated from their critical volumes by the corresponding states relations, $V_0 = 0.31V_c$.

We stated in refs. 2–5 that diffusivity of different solutes in the solvent—except for the "quantum gases"—appeared to

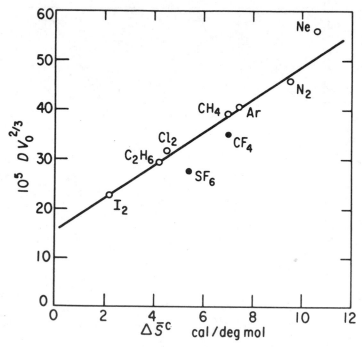

FIG. 3.   Relation of diffusivity to molecular cross-section and to configuration entropy.

vary inversely with their molecular cross-sections as measured by the squares of their molecular diameters, but later, because reported σ-values for the gas do not agree, we substituted the $2/3$ power of either boiling points or critical molal volumes. We now use values of V_0 obtained from fluidity, and give in the table values for $10^5 DV_0^{2/3}$ at $25°$. The wider range of D values now available shows that this product is far from uniform and that an additional parameter is required.

In ref. 11 we wrote:

In the case of solutions so dilute as these here considered, these changes represent what happens in the immediate neighborhood of solute gas molecules. These have small attractive forces. . . .but the same kinetic energies as the molecules of the solvent, hence the latter gain added volume and freedom of motion such as they would gain at the surface of a bubble.

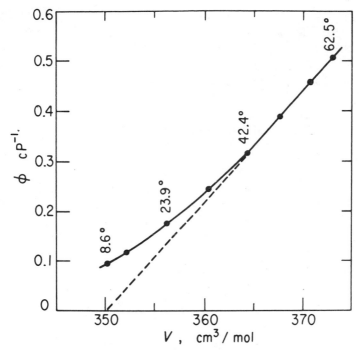

FIG. 4. Variation of fluidity of $(C_4F_5)_3N$ with molal volume and temperature.

We illustrated the magnitude of this expansion by figures for the percent excess of the partial molal volumes of C_2H_6, CH_4, and N_2 in CCl_4 over their molal volumes at their boiling points, which are, respectively: 33.5, 51.8, and 67.5%. This added freedom of motion contributes notably to diffusivity.

Many years of investigating the solubility and entropy of solution of nonreactive gases in nonpolar liquids have yielded information about their entropy of solution in CCl_4 that is applicable to diffusivity. In Fig. 2 we plot solubility at 25° expressed as $-R \ln x$ on the horizontal axis; on the left-hand vertical axis are values of partial molal entropy of solution from gas at 1 atmosphere into saturated solution. We see that points fall close to a straight line that intersects the vertical

TABLE 1. *Diffusion coefficients, 10^5 D cm^2/sec, of gases in CCl_4; V_0, intrinsic molal volume of gas at zero fluidity, cm^3/mol; $\Delta\bar{S}^c$, cal/deg-mol, excess of partial molal entropy of solution over entropy of dilution at $25°$.*

| | Temperature | | | $V_0^{2/3}$ | $10^5 DV_0^{2/3}$ | $\Delta\bar{S}^c$ |
	5°	25°	35°			
H_2	7.64	12.0	—	8.3	100	9.6
Ne	5.94	9.44	—	5.90	55.7	10.6
Ar	3.10	4.85	—	8.43	40.9	7.4
N_2	3.15	4.76*	—	9.65	46.0	8.74
CH_4	2.54	3.86	—	10.4	39.0	7.0
C_2H_6	—	2.36	2.79	12.6	29.6	4.0
Cl_2	—	2.88	3.34	11.0	31.7	4.2
I_2	—	1.50	1.80	14.8	22.2	2.1
CF_4	1.65	2.50	—	12.8	36.0	7.2
SF_6	—	1.78	2.07	16.0	28.4	4.7
CCl_4	—	1.30	1.79	20.2	26.2	—

* Recalculated from data in ref 4. —, not done.

axis at -21.0 cal/deg·mol. As this line is descended, the successive solutes become more and more like the solvent until they coalesce to a single substance with a boiling point of $25°$ and a molal entropy of evaporation of $21°$ cal/deg. mol. This value is a base for calculating an *excess partial molal entropy* ΔS^E. We split this into *dilution entropy*, $-R \ln X$ and *configuration entropy*, $\Delta\bar{S}^c$, as shown on the right-hand vertical axis. This is presented in detail in ref. 11. The values of $\Delta\bar{S}^c$ for all solutes except Cl_2 and I_2 are from that paper. The value for Cl_2 was calculated from data by Taylor and Hildebrand (12). We calculated the value for I_2 from the excess of its partial molal volume, 66.7 cm^3, over its liquid molal volume (13) extrapolated to $25°$, 59 cm^3, using the equation $\Delta\bar{S}^c = (\bar{V} - V_0)(\partial S_1/\partial V)_T$ (see ref. 9, p. 81); $(\partial S_1/\partial V)_T$ for CCl_4 at $25°$ $= 0.27$ cal/cm^3·deg.

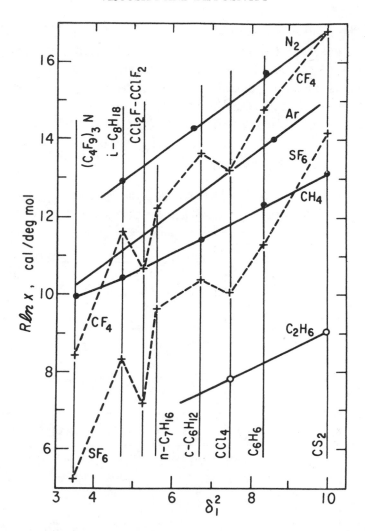

FIG. 5. Irregular relations of solubility of CF_4 and SF_6 in liquids with different solubility parameters.

Values of $10^5 D V_0^{2/3}$ are plotted against those of $\Delta \bar{S}^c$ in Fig. 3. One sees that diffusion is indeed enhanced by the expansion and accompanying additional molecular freedom about solute molecules with low attractive potential.

The position of the point for Ne, 5 units above the line for other gases, and the point for H_2, 52 units above the line, are nearly in the same ratio as the squares of their de Boer–Michels quantum mechanical parameters; 0.35 and 3.0, confirming the findings published in refs. 3 and 4.

The displacement of the points for CF_4 and SF_6 is not surprising in view of the irregularity of their solubility in different solvents, illustrated in Fig. 4.

TABLE 2.   *Viscosity of $(C_4F_9)N$ (centipoise) and molal volume, V (cm^-) at temperatures $t°$*

$t°$	η	V	$t°$	η	V
8.60	10.378	350.19	42.40	3.151	364.23
13.31	8.478	352.05	50.23	2.595	367.71
23.91	5.619	356.38	58.10	2.170	371.39
33.38	4.097	360.36	62.50	1.965	373.26

TABLE 3.   *Diffusion coefficients, 10^5D (cm/sec) of gases in $(C_4F_9)_3 N$*

	5°	10°	15°	25°	35°	40°
H_2	6.59	6.98	—	8.21*	—	—
				8.30*	—	
Ne	4.38	—	5.19	6.34*	7.89	—
				6.37*		
Ar	2.20	—	2.50	6.21	4.20	—
CH_4	—	—	—	3.05	—	—
	—	—	—	3.18*	—	—
C_2H_6	—	—	1.19	1.85	2.60	2.96
CF_4	1.13	—	1.50	2.15	3.04	—
			1.48	2.21*	3.07	—
					3.01	—

* Powell, R. J. & Hildebrand, J. H. (1971) *J. Chem. Phys.* **55,** 4715–4716.

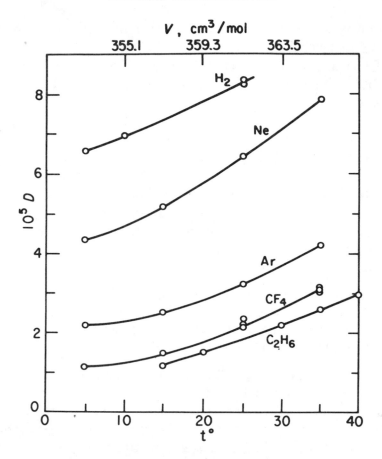

FIG. 6.   Diffusivity of gases in $(C_4F_9)_3N$ at 25°.

DIFFUSIVITY IN $(C_4F_9)_3N$

This solvent differs strongly from CCl_4 in molal volume, 360 compared with 97 cm³, and in the symmetry of their molecules; those of the amine have three bulky arms. The line of fluidity against molal volume is straight for CCl_4, but strongly curved at smaller volumes for $(C_4F_9)_3N$. We determined the viscosity of our sample of $(C_4F_9)_3N$, with results shown in Table 2, and plotted a fluidity, $\phi = 1/\eta$, in Fig. 5. They agree well with data furnished by the 3M Co. The strong curvature was expected from findings with other liquids with unsymmetrical molecules, discussed in ref. 13.

Our values for diffusivity of gases in $(C_4F_9)_3N$ are shown in Table 3 and plotted in Fig. 6. There are striking differences between this plot and that for diffusivity in CCl_4. The lines are not straight and do not converge to zero at a common temperature. H_2 and Ne continue to diffuse at temperatures where the solvent becomes highly viscous; their small molecules evidently find channels for diffusion between the more or less interlocked cogwheel-like molecules of the solvent.

In the superb book on *Transport Phenomena*, by R. B. Bird, W. E. Stewart, and E. N. Lightfoot (Wiley, 1960), the authors, after reviewing attempts that have been made to form valid theories of diffusivity, write (p. 515):

If the reader has by now concluded that little is known about the prediction of dense gas and liquid diffusivities, he is correct. There is an urgent need for experimental measurements, both for their own value and for development of future theories.

By selecting gases as diffusants in a simple liquid, we have widely varied parameters and revealed the effects of molal volume and of configuration entropy. By varying temperature, we have shown that diffusivity is a function of $T - T_0$, not T, and laid a realistic foundation for future studies free from assumptions of activation.

Acknowledgment is made to the donors of The Petroleum Research Fund, administered by the American Chemical Society, for support of this research, and to the Minnesota Mining and Manufacturing Co. for the supply of perfluorotributyl amine.

1. Haycock, E. W., Alder, B. J. & Hildebrand, J. H. (1953) *J. Chem. Phys.* 21, 1601–1604.
2. Watts, H., Alder, B. J. & Hildebrand, J. H. (1955) *J. Chem. Phys.* **23**, 659–661.
3. Ross, M. & Hildebrand, J. H. (1967) *J. Chem. Phys.* **40**, 2397–2399.

4. Nakanishi, K., Voight, E. M. & Hildebrand, J. H. (1965) *J. Chem. Phys.* **42**, 1860–1863.
5. Powell, R. J. & Hildebrand, J. H. (1971) *J. Chem. Phys.* **55**, 4715–4716.
6. Hildebrand J. H. (1971) *Science* **174**, 490–493.
7. Clegg, G. T. & Tehrani, M. A. (1973) *J. Chem. Eng. Data*, **18**, 59–60.
8. Collings, A. F. & Mills, R. (1970) *Trans. Faraday Soc.* **66**, 2761–2766.
9. Hildebrand, J. H. & Lamoreaux, R. H. (1972) *Proc. Nat. Acad. Sci. USA* **69**, 3428–3431.
10. Hildebrand, J. H., Prausnitz, J. M. & Scott, R. L. (1970) in *Regular and Related Solutions* (Van Nostrand Reinhold, New York), p. 118.
11. Hildebrand, J. H. & Lamoreaux, R. H. (1974) *Chem. Eng. Fundament.* **13**, 110–114.
12. Taylor, N. W. & Hildebrand, J. H. (1923) *J. Amer. Chem. Soc.* **45**, 682–694.
13. Shinoda, K. & Hildebrand, J. H. (1958) *J. Phys. Chem.* **62**, 295–296.

9

Diffusivity of Methane in a Mixture of CCl_4 and c-$C_6F_{11}C_2F_5$ of the Critical Composition in the Region above the Temperature of Separation

(liquid–liquid critical mixture)

J. H. HILDEBRAND AND R. H. LAMOREAUX

Department of Chemistry, University of California, Berkeley, Calif. 94720

Contributed by J. H. Hildebrand, June 18, 1974

ABSTRACT   The diffusivity of CH_4 in a mixture of CCl_4 and c-$C_6F_{11}C_2F_5$ of the critical composition in the region of temperature close to that of unmixing, decreases as in a homogeneous liquid from 36° to about 32°. It then passes through a minimum of $10^5D \approx 4.15$ cm²/sec at about 27.5°, then rises to $10^5D = 4.61$ at 25.00°, then steeply to 6.36 cm²/sec in the further drop of only 0.3° to 24.71°.

The interesting region of liquid–liquid mixtures just above their maximum consolute point has been investigated in several ways. Zimm (1) in 1950 found that the turbidity of a critical mixture of CCl_4 and c-$C_6F_{11}CF_3$ decreased rapidly from a value of >20 at 28.31° to 4.0 at 28.34° to 0.15 at 28.9°, to 0.026 at 30.9. Jura, Fraga, Maki, and Hildebrand (2) reported the variation with temperature of the critical

Reprinted from
Proc. Nat. Acad. Sci. USA
Vol. 71, No. 10, pp. 3800–3801, October 1974

volume of a mixture of i-C_8H_{18} and C_7F_{16} from well above the consolute temperature to well below. They also found a λ-type cusp in heat capacity with variations in composition. This was confirmed later with greater precision by Schmidt, Jura, and Hildebrand (3). Sound attenuation in aniline + hexane was reported by Chynoweth and Schneider (4). Reed and Taylor (5) found a strong cusp in viscosity of i-C_8H_{18} + C_7F_{16} at 23.7°, the critical temperature, which becomes weak at 30° and has disappeared at 35°. Hildebrand, Alder, Beams and Dixon (6) found that the consolute temperatures of two liquid pairs, CCl_4 + $C_6F_{11}CF_3$ and i-C_8H_{18} + C_7F_{16}, were raised by as much as 10° in a centrifuge accelerated to 10^8 cm/sec², due to the hydrostatic effect, and that the second system, whose components differ in density by 1.02, exhibit an added sedimentation effect of +2°.

We decided that it would be worthwhile to investigate the diffusivity of a solute gas in this region, a phenomenon we did not feel competent to predict with any certainty.

Experimental

We selected CCl_4 and c-$C_6F_{11}C_2F_5$ as the liquid-liquid mixture and CH_4 as the diffusant. The purity of each was quite adequate. The density of the fluorocarbon was 1.8254 g/cm³ at 20°. The critical mixture contains 60 volume percent of CCl_4, and the critical temperature is 24.70°.

Diffusivity was measured by the method and apparatus

TABLE 1. *Values of diffusivity, 10^5D, in cm²/sec at decreasing temperatures to about 0.02° above the critical temperature*

$t°$	10^5D	$t°$	10^5D
35.80	5.53	25.00	4.61
32.88	4.80	24.91	4.88
30.00	4.27	24.81	5.40
29.82	4.27	24.73	6.28
27.10	4.16	24.71	6.36

described by Hildebrand and Lamoreaux (7) with the sub-
stitution of a thin diaphragm of sintered bronze to shorten
diffusion times.

Results

Results are shown in Table 1 and plotted in Fig. 1. One sees
that diffusivity of methane decreases from 36°, at first nearly
linearly with temperature and as is usual in a homogeneous
liquid, but that it goes through a minimum, and in the final
0.29° rises sharply from 4.61 to 6.36 cm²/sec.

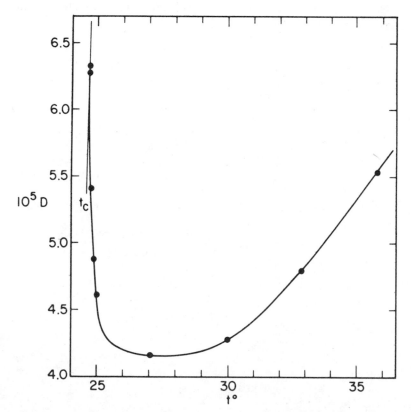

FIG. 1.   Diffusivity of CH_4 in a mixture of CCl_4 and $c\text{-}C_6F_{11}C_2F_5$
of the liquid-liquid critical composition from 36° to the critical
temperature.

The diffusivity of CH_4 in pure CCl_4 at 27.0° is only 3.84 cm^2/sec.

Anyone who had inferred that the high viscosity in this region would decrease diffusivity would have been completely wrong.

It is instructive to observe the mixture as its temperature is slowly decreased. Above $t_c + 5$, it is clear and colorless. A faint blue appears about 3° above t_c, depending upon the difference between refractive indices; this color rapidly deepens, and the mixture becomes turbid. At t_c the liquid suddenly becomes an unstable, coarse emulsion of colorless phases which settle into clear liquids, separated by a flat meniscus.

Acknowledgement is made to the donors of the Petroleum Research Fund, administered by the American Chemical Society, for the support of this research.

1. Zimm, B. H. (1950) *J. Phys. Coll. Chem.*, **54**, 1306–1317.
2. Jura, G., Fraga, G., Maki, G. & Hildebrand, J. H. (1953) *Proc. Nat. Acad. Sci. USA* **39**, 19–23.
3. Schmidt, H., Jura, G. & Hildebrand, J. H. (1959) *J. Phys. Chem.* **63**, 297–299.
4. Chynoweth, A. G. & Schneider, W. G. (1951) *J. Chem. Phys.* **19**, 1607.
5. Reed, T. M. & Taylor, T. E. (1959) *J. Phys. Chem.* **63**, 58–67.
6. Hildebrand, J. H., Alder, B. J., Beams, J. & Dixon, H. M. (1954) *J. Phys. Chem.* **58**, 577–579.
7. Hildebrand, J. H. & Lamoreaux, R. H. (1974) *Proc. Nat. Acad. Sci. USA* **71**, 3321–3324.

10

Kinetic Theory of Viscosity of Compressed Fluids

J. H. HILDEBRAND

Department of Chemistry, University of California,
Berkeley, Calif. 94720

Contributed by J. H. Hildebrand, March 10, 1975

ABSTRACT   The viscosity of a compressed gas is a function of three variables: (*1*) the degree of crowding of the molecules; (*2*) their capacity, by reason of softness, flexibility, or rotational inertia, to absorb the vector momentum applied to cause flow; (*3*) the resistance to this vector momentum offered by the randomly oriented thermal momenta, which becomes significant when the liquid expands sufficiently to permit molecular mean free paths between binary collisions to be long enough for thermal momenta to acquire fractions of their thermal momentum in free space.

The fluidity ϕ of simple liquids obeys the linear equation $\phi = B(V - V_0)/V_0$ and its viscosity is, therefore, $\eta_a = V_0/B(V - V_0)$; this accounts for components *1* and *2*. The contribution of random thermal momenta, *3*, obeys the equation, $\eta_b = \eta_0(1 - V_t/V)$. η_0 is the viscosity of the dilute gas; V_t is the molal volume at which the thermal contribution begins. The total momentum, $\eta = \eta_a + \eta_b$.

Values of η_0 vary linearly with $T^{1/2}$. Values of V_t are related to heat capacities.

Reprinted from
Proc. Nat. Acad. Sci. USA
Vol. 72, No. 5, pp. 1970–1972, May 1975

Hildebrand and Lamoreaux (1) in a recent paper titled "Viscosity Along Continuous Paths Between Liquid and Gas," examined data for carbon dioxide and propane; this paper reports an extension of the study to data for argon, methane, ethane, normal butane, and carbon dioxide. It begins, as before, with the liquid range; where, as we (2, 3) have shown by many examples, fluidity, ϕ, the reciprocal of viscosity, η, is a linear function of molal volume, V, at all pressures and temperatures up to volumes well above boiling point volumes. The equation is

$$\phi = B(V - V_0)/V_0 \tag{1}$$

Values of B and V_0 are easily determined from plots of ϕ against V. Viscosity of a liquid therefore conforms closely to the equation

$$\eta_a = V_0/B(V - V_0) \tag{2}$$

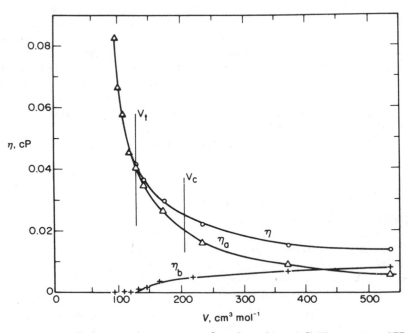

FIG. 1.   Relations between η, the viscosity of C_3H_8 at 410.9°K, and its components η_a and η_b.

We add the subscript a because an additional source of viscosity, designated η_b, enters as V approaches the critical molal volumes, V_c, where molecules are no longer in the potential field of many near neighbors and acquire mean free paths between discrete collisions and thermal momenta, which are randomly oriented and oppose the vector momentum applied to cause the flow. This added source of viscosity, η_b, increases from zero at V_t, the molal volume at which the thermal contribution begins, to its value in the dilute gas, η_0, by the equation

$$\eta_b = \eta_0(1 - V_t/V) \qquad [3]$$

Values of V_0 and B for the gases here considered, shown in Table 1, were obtained from plots of ϕ against V in the liquid region. The sources of the data are given in the list of references. Values of V_t were derived from plots like the one in Fig. 1 for C_3H_8 at $410.9°K$. In the earlier paper, dealing only with C_3H_8 and CO_2, we had inferred that η_b begins, at least for

TABLE 1. *Values of parameters for molecules*

Molecule	V_0	V_t	V_c	B	C_p
Ar	24.8	22	74.6	23.6	5.0
CH_4	32.0	40	99.4	35.6	8.5
C_2H_6	44.6	64	148	25.8	12.6
C_3H_8	61.5	130	205	22.0	17.6
n-C_4H_{10}	78.5	230	255	21.8	23.6
CO_2	28.7	60	94	15.7	8.9

Volumes, $cm^3\ mol^{-1}$: where flow begins, V_0; where thermal momenta become effective, V_t; critical, V_c. B is constant of proportionality between fluidity and volume in Eq. 1, reciprocal centipoise (g^{-1} msec). C_p is heat capacity of dilute gas, at $25°C$, in cal $mol^{-1}\ °K^{-1}$ (1 cal = 4.184 J).

Sources of data: ref. 4; Ar, refs. 5–8; CH_4, refs. 9–11; C_2H_6, ref. 10; C_3H_8, ref. 12; n-C_4H_{10}, ref. 13; CO_2, refs. 14 and 15.

these fluids, at one half the critical volume, but one sees now that the values of V_t/V_c in Table 1 range from 0.30 for Ar to 0.90 for n-C_4H_{10}. Furthermore, we then wrote $1 - (V_c/2V)^{2/3}$ for the function of V in Eq. 3, taking the $2/3$ from the equation for mean free paths. The simplification shown in Eq. 3 is gratifying. Values of V_t can also be calculated from η_b and η_0, at a single value of V.

The viscosity of a dilute gas increases with temperature, since the thermal momentum of a hard sphere is proportional

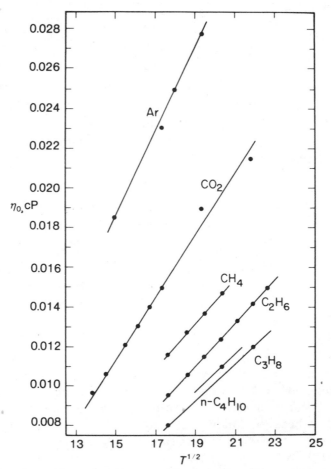

FIG. 2.   The viscosity of a dilute gas is a linear function of the square root of the temperature.

TABLE 2. *Viscosity, η_0, in centipoise, of dilute gases at different temperatures, °K*

T	CH_4	C_2H_6	C_3H_3	$n\text{-}C_4H_{10}$	T	Ar	T	CO_2
311	0.0116	0.0096	—	0.0080	223	0.0185	200	0.0102
344	0.0127	0.0106	—	—	298	0.0231	220	0.0111
378	0.0137	0.0115	—	—	323	0.0250	240	0.0121
411	0.0147	0.0124	0.0110	—	373	0.0278	260	0.0130
444	—	0.0138	—	—	473	0.0330	280	0.0140
478	—	0.0142	—	0.0120			300	0.0150
511	—	0.0150	—	—				

to $(mT)^{1/2}$. It is well known, however, that the viscosity of a gas of soft molecules is not strictly proportional to $T^{1/2}$, [compare Sir James Jeans (16)], but it is linear, as illustrated in Fig. 2, which plots the data given in Table 2.

Fig. 3 presents the evidence for the validity of Eq. **3**. If values of η_b/η_0 are plotted against $1 - V_t/V$, consistent measurements made at different temperatures are expected to fall on a single curve for each substance. The points represent values of $(\eta - \eta_a)/\eta_0$ obtained from measurements

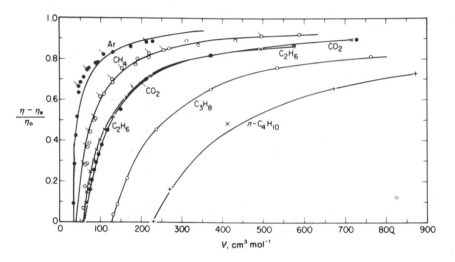

FIG. 3. Test of the relation $\eta_b/\eta_0 = 1 - V_t/V$.

of liquid, compressed and dilute gas, and may fairly be designated "experimental"; the curves, however, were drawn from Eq. **3**, and are hence "theoretical." The points for Ar include some at 323 and 373°K; those for CH_4 are at 311 and 344°K; those for C_3H_8 are at 411 and 511°K. For the latter, η_0 was obtained by extrapolation on the plot shown in Fig. 2.

The shapes of the curves in Fig. 3 reveal information about differences between molecular collisions of hard or

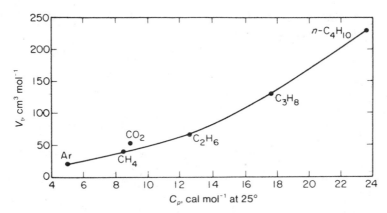

FIG. 4. Relation between V_t, the volume at which random molecular thermal momentum begins to contribute to viscosity, and C_p, the molar heat capacity of the dilute gas.

soft molecules; these differences have received little attention. Molecules of Ar evidently have considerable thermal momentum even in the liquid; the ratio V_t/V_0 is 0.95 for Ar, 1.25 for CH_4, 2.93 for n-C_4H_{10}. Values of V_t are closely related to the molar heat capacities of the dilute gases as illustrated in Fig. 4.

I am indebted to my colleagues, B. J. Alder and J. M. Prausnitz for their interest and criticism.

1. Hildebrand, J. H. & Lamoreaux, R. H. (1974) *Physica* **74**, 416–422.
2. Hildebrand, J. H. & Lamoreaux, R. H. (1972) *Proc. Nat. Acad. Sci. USA* **69**, 3428–3431.
3. Hildebrand, J. H. & Lamoreaux, R. H. (1973) *J. Phys. Chem.* **71**, 1471–1473.
4. Rossini, F. D., and others (1971) *American Petroleum Institute Research Project no. 44* (Carnegie Press, Pittsburgh, Pa.).
5. Hayeres, W. M. (1973) *Physica* **67**, 440–470.
6. Michels, A., Botzen, H. & Schuurman, W. (1945) *Physica* **20**, 1141–1148.
7. Reynes, E. G. & Thodos, G. (1964) *Physica* **30**, 1529.
8. Kestin, J. & Whitelaw (1963) *Physica* **29**, 335–356.

9. Matthews, C. S. & Hurd, C. O. (1946) *Trans. Am. Inst. Chem. Eng.* **42**, 55–58.
10. Eakin, B. E., Starling, K. G. & Dolan, J. P. (1962) *J. Chem. Eng. Data* **7**, 33–36.
11. Sage, B. H. & Lacey, W. N. (1950) *Thermodynamic Properties of the Lighter Hydrocarbons and Nitrogen* (American Petroleum Inst. New York).
12. Starling, K. E., Eakin, R. E. & Ellington, R. T. (1960) *AIChE J.* **6**, 438–442.
13. Dolan, J. P., Starling, K. E., Lee, A. L. & Ellington, R. T. (1963) *J. Chem. Eng. Data* **8**, 396–399.
14. Michels, A., Botzen, A. & Schuurman, W. (1957) *Physica* **23**, 95–102.
15. Johnson, H. L. & McKloskey (1940) *J. Phys. Chem.* **44**, 1039–1058.
16. Jeans, J. (1953) in *An Introduction to the Kinetic Theory of Gases* (Cambridge Univ. Press, New York), p. 171.

11

Viscosity of liquid metals:
An interpretation

(fluidity of liquid metals/solubility parameters)

J. H. HILDEBRAND AND R. H. LAMOREAUX

Department of Chemistry, University of California, Berkeley, Calif. 94720

Contributed by Joel H. Hildebrand, January 9, 1976

ABSTRACT   The fluidity of liquid metals, like that of simple nonmetallic liquids, is a linear function of the ratio of unoccupied volume to intrinsic volume over long ranges as expressed by the equation $\phi = B[(V/V_0) - 1]$. Values of V_0 obtained by extrapolating to $\phi = 0$ agree well with molal volumes of compact crystals at 20°C calculated from densities.

Values of the ratio $\phi/[(V/V_0) - 1]$ range from 27.0 for Na to 1.55 reciprocal centipoise for Ni. Viscosities at ratios of expansion $V/V_0 = 1.10$ vary linearly with squares of solubility parameters $\Delta E^v/V_0$, where ΔE^v is molal energy of vaporization at the melting point. Viscosities at 10% expansion range from 0.037 cP for Na to 5.38 cP for Co, with some divergence for metals with values of η in the neighborhood of unity. The good agreement in the case of the transition metals Cu, Fe, Co, and Ni we attribute to distribution of vector momentum among quasi-chemical bonds between d-electrons and vacant orbitals.

Reprinted from

Proc. Nat. Acad. Sci. USA
Vol. 73, No. 4, pp. 988–989, April 1976
Chemistry

We have shown in a series of papers (1–5) that fluidity ϕ (the reciprocal of viscosity, η) of simple liquids is a linear function of their molal volumes V over considerable ranges of temperature and pressure as expressed by the simple equation:

$$\phi = B[(V/V_0) - 1] \qquad [1]$$

The intercept of the ϕ–V line at $\phi = 0$ is the intrinsic, close-packed molal volume. The proportionality constant B is largest for symmetric, elastic molecules that do not transpose vector momentum into rotational or vibrational energy. The purpose of this paper is to examine the applicability of these concepts to an understanding of the viscosity of liquid metals.

Data on the viscosities of liquid metals over ranges of temperature are given in the Landolt-Börnstein Tables (6). To calculate molal volumes over ranges of temperature we used the extensive compilation by Cowley (7).

We obtained lines that are quite straight, except at high temperatures in a few cases to be discussed. Extrapolation to V_0 at $\phi = 0$ involves a considerable interval in the lines for Mg and Ca, consequently the precision of a value of V_0 is limited by the length of this interval, the range of measurements, and the scatter of points. Our values of V_0 are listed in Table 1. They differ a little from corresponding values obtained by Wittenberg and DeWitt (8) in an important paper on changes of volume on melting, mainly because we used a different source for the dependence of volume upon temperature.

In the second column of figures in Table 1 we give values of the molal volumes of the solid metals at 20°C, calculated from densities given in the American Physical Society *Handbook of Physics* (8). These values vary, in some cases by as much as 2%, depending upon the method of preparing the sample or whether calculated from x-ray-determined dimensions. We discussed in reference 2 the conditions under which V_0 and V_s might have the same value. It is striking

Table 1. Values of parameters

		V_0	V_s	$\dfrac{\phi}{V/V_0 - 1}$	$\dfrac{\Delta E^v}{V_0}$	η
1	Li	12.4	13.0	14.5	2.9	0.07
2	Na	24.5	24.1	27.0	0.56	0.04
3	K	44.2	45.4	27.0	0.45	0.04
4	Rb	55.3	55.8	27.0	—	—
5	Mg	~13	14.0	~3.5	2.3	—
6	Ca	~23	26.3	~3.1	1.6	—
7	Al	10.7	10.0	11.0	7.2	0.91
8	Fe	7.05	7.05	2.04	12.3	4.9
9	Co	6.80	6.80	1.86	12.7	5.4
10	Ni	6.70	6.70	1.55	13.8	6.5
11	Cu	7.10	7.14	2.3	9.8	4.4
12	Zn	9.50	9.45	5.7	2.7	1.8
13	Ag	11.0	10.28	3.9	6.1	2.56
14	Cd	13.0	13.0	5.0	1.7	2.0
15	Hg	14.1	14.14	11.9	2.0	0.84
16	Ga	11.1	11.8	15.0	—	0.67
17	In	15.7	15.8	12.5	3.5	0.80
18	Sn	16.4	16.3	13.0	4.1	0.77
19	Pb	18.8	18.3	8.8	2.3	1.14
20	Sb	17.6	18.4	10.6	—	0.94
21	Bi	19.6	21.4	9.8	4.5	1.0

ϕ is fluidity in reciprocal centipoise, cP^{-1} ($1 P = 1\,g\,cm^{-1}\,sec^{-1}$). V is volume in $cm^3\,mol^{-1}$. V_0 is its value at $\phi = 0$; V_s is its value for the solid at 20°C. ΔE^v is energy of vaporization in $kcal\,mol^{-1}$ (1 kcal = 4.184 kJ) of the liquid metal at its melting point. η is viscosity, in cP, of liquid metal when $V/V_0 = 1.10$.

that this is the case with many of the metals in Table 1. In the cases of Ga, Sb, and Bi, which expand on freezing, V_s is, of course, larger than V_0. Their fields of force are not symmetrical; Ga crystals are orthorhombic and those of Sb and Bi are rhombohedral.

The purpose of this paper is not to consider change of volume on freezing but only to show that the parameter V_0 derived from fluidity at high temperatures represents a real physical quantity, the intrinsic volume of a mole of close-packed, highly symmetrical molecules, appropriate for our purpose of dealing with the viscosity of liquids.

In order to study the factors that are embodied in the constant B of Eq. **1,** we have plotted in Fig. 1 values of ϕ versus $(V/V_0) - 1$ for each metal. Values of the slopes, given in Table 1, give the fluidities at equal expansions over intrinsic volumes. Their enormous range, from 27 to 1.55, invites theoretical analysis. Lines for the non-metals, Ne, CH_4, $SiCl_4$, and Cl_2, have been included for comparison. Their slopes decrease with liquid heat capacities (5). It is striking that the points for alkali metals other than Li all fall on the same line, the steepest with fluidity far higher than that of any

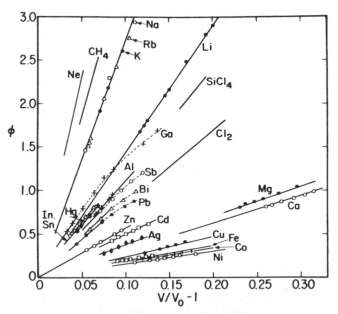

FIG. 1.   Fluidity of liquid metals, ϕ in cP^{-1}, against ratios of free volume to occupied volume.

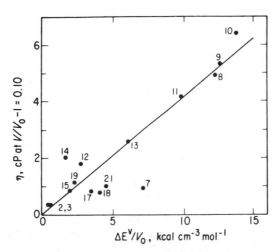

FIG. 2.   Viscosity, η, cP, at liquid expansion $V/V_0 - 1 = 0.100$ against cohesion expressed by molal energy of vaporization per cm^3 at the melting point, kcal cm^{-3} mol^{-1}. Numbers relate the points to the elements in Table 1.

other metals and not much less than that of CH_4 at the same degree of expansion. This seems consistent with the fact that the atom of each of these metals possesses a single valence electron in symmetrical s-orbital.

Let us view the same data from the standpoint of viscosity, all at the expansion 1.10 V/V_0, plotted as ordinates in Fig. 2. As abscissae are plotted values of molal energy of vaporization at melting points divided by V_0, $\Delta E^v/V_0$. We have subtracted RT from values of heats of vaporization at melting points from *Selected Values of the Thermodynamic Properties of the Elements* (9). $\Delta E^v/V_0$ is the square of the "solubility parameter," δ, now widely used as a measure of cohesion. Applied to liquid metals, it includes not only purely physical, London dispersion forces, but also the much larger quasi-chemical forces between the atoms with d-electrons and vacant orbitals in the transition elements, notably Cu, Fe, Co, and Ni.

The group of metals with values of $\Delta E^v/V_0$ between 2 and 4 kcal cm^{-3} mol^{-1} are considerably more viscous than the alkali metals Na to Cs, indicating probability of some formation of hybrid interatomic orbitals. Our present quantitative knowledge of forces in liquid Sn, In, Hg, Pb, and Al seems insufficient to predict their absorbtivity for vector momentum. Perhaps viscosity has more to teach than to learn about these interactions.

We thank colleagues Leo Brewer, Kenneth Pitzer, and John Prausnitz for helpful discussions of this subject.

1. Hildebrand, J. H. (1971) "Motions of molecules in liquids," *Science* **174**, 490–493.
2. Hildebrand, J. H. & Lamoreaux, R. H. (1972) "Fluidity: A general theory," *Proc. Nat. Acad. Sci. USA* **69**, 3428–3431.
3. Alder, B. J. & Hildebrand, J. H. (1973) "Activation energy: Not involved in transport processes in liquids," *Ind. Eng. Chem. Fundam* **12**, 387–388.
4. Hildebrand, J. H. & Lamoreaux, R. H. (1973) "Viscosity along continuous paths between liquid and gas," *Physica* **74**, 416–422.
5. Hildebrand, J. H. (1975) "Kinetic theory of viscosity of compressed fluids," *Proc. Nat. Acad. Sci. USA* **72**, 1970–1972.
6. Landolt-Börnstein *Tabellen* (Springer-Verlag, Berlin), II 5a, p. 124 ff.
7. Crawley, A. F. (1974) "Densities of liquid metals and alloys," *Int. Metall. Rev.* (Department of Energy, Mines and Resources, Ottawa, Canada), pp. 32–40.
8. Wittenberg, L. J. & DeWitt, R. (1972) "Volume contraction during melting; emphasis on lanthanide and actinide elements," *J. Chem. Phys.* **56**, 4526–4533.
9. Hultgren, R. (ed.) (1973) *Selected Values of the Thermodynamic Properties of the Elements* (Wiley, New York).

12

Viscosity of dilute gases and vapors

(fluidity/molal volume/polyatomic molecules)

J. H. HILDEBRAND

Department of Chemistry, University of California, Berkeley, Calif. 94720

Contributed by Joel H. Hildebrand, October 7, 1976

ABSTRACT   The well-known formula for calculating the viscosity of hard sphere gases, $\eta_o = 2(mkT)^{1/2}/3\pi^{3/2}\sigma^2$, where σ^2 is molecular cross section, is altered to $\eta_o = K(MT)^{1/2}/V_t^{2/3}$, where V_t is the molal volume of a liquid that has expanded sufficiently to permit mean free paths long enough to have significant fractions of the random thermal momenta of molecules in free flight and M is molal mass.

This formulation places upon a single straight line points for all nonpolar molecules, mono- and polyatomic molecules alike, over long ranges of temperature.

Kinetic theory has yielded the equation

$$\eta_o = 2(mkT)^{1/2}/3\pi^{3/2}\sigma^2 \qquad [1]$$

for the viscosity of a dilute gas of hard sphere molecules. m is molecular mass, σ is molecular diameter, k is the Boltzmann constant, and T is the thermodynamic temperature. η_o is the vector momentum applied to produce flow; $(mkT)^{1/2}$ repre-

Reprinted from
Proc. Natl. Acad. Sci. USA
Vol. 73, No. 12, pp. 4302–4303, December 1976
Physics

sents the randomly oriented thermal momenta that produce viscosity.

Sigma is not a satisfactory parameter because it is calculated either from Eq. 1 or by some method that assumes an expression for pair-potentials, such as the familiar 6–12 potential of Lennard-Jones, not valid for polyatomic molecules (see ref. 1). Accordingly, I substitute for σ^2 values of $V_o^{2/3}$, where V_o is intrinsic molar volume of a liquid obtained from the linear relation between molal volume V and fluidity, ϕ in the equation (2, 3)

$$\phi = B(V - V_o)/V_o \qquad [2]$$

Eq. 1 can then be written

$$\eta_o = K(MT)^{1/2}/V_o^{2/3} \qquad [3]$$

where M is the molal mass. We (4) applied this to data for propane and CO_2, with V_o values 61.0 and 28.4 cm^3 mol^{-1}, respectively, and obtained $K = 1.36 \times 10^{-3}$ for propane and 1.38×10^{-3} for CO_2.

We found that the linear relation between ϕ and V holds to large pressures and smaller volumes, also to volumes approximately of half of critical volumes, but as critical volumes are approached, the straight line splits into isotherms that bend toward the horizontal and approach constant fluidity at high dilution. Viewed from the standpoint of viscosity, the reciprocal of fluidity, in Fig. 1, for propane, reproduced from refs. 4 and 5, the split occurs at the volume V_t, where η_a is the contribution of crowding to viscosity calculated by Eq. 2. As η_a approaches zero, η_b increases to η_o, beginning at V_t. The shape of η_b is given by the equation

$$\eta_b = \eta_o(1 - V_t/V). \qquad [4]$$

In order to obtain precise values of V_t from plots of experimental points of either fluidity or viscosity, there must be an

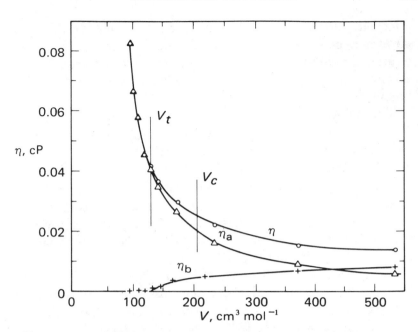

FIG. 1. Relation between the viscosity of propane at 410.9 K and its components η_a and η_b. cP is centipoise; $1\ P = 1\ g\ cm^{-1}\ sec^{-1}$.

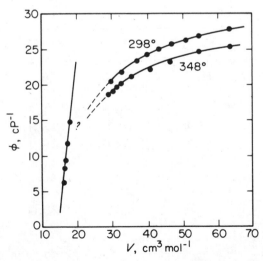

FIG. 2. Fluidity of neon at 298 K and 348 K, showing gap in data between 20 and 30 $cm^3\ mol^{-1}$.

adequate number of points. This is not the case for all of the gases here considered. The available data for neon are plotted as an example in Fig. 2, where there is a gap between 20 and 30 cm³ mol⁻¹ How I have handled such cases will be described below.

This paper deals with the nine gases listed in Table 1 and in Fig. 3, where values of viscosity over temperature ranges of from 100° to 200° are plotted against $(MT)^{1/2}$. I invite attention, first, to the straightness of all the lines over their ranges of temperature, as demanded by kinetic theory.

The ratios of V_t/V_o given in Table 1 increase strongly with molecular atomicity and consequent capacity to absorb vector momentum instead of only distributing it from one molecule to another, as with hard spheres.

In Fig. 4, the viscosities are plotted against $(MT)^{1/2}/V_t^{2/3}$ for gases here considered. Only two values at the extremes are plotted to avoid crowding. In cases where data do not fix values accurately, values of $V_t^{2/3}$ have been adjusted to conform to the single line yielded by cases more clearly indicated. In the cases of bromine and chloroform I simply picked a value of

Table 1. Molal volumes, cm³

Gas	V_o	V_t	$V_t^{2/3}$	V_t/V_o
Ne	14.0	16.6	6.2	1.19
CH_4	32.0	39.3	11.4	1.23
Ar	24.5	39.5	11.5	1.49
CO_2	23.4	60.0	15.3	2.11
C_2H_6	44.6	64.0	16.0	1.43
Br_2	45.8	72.5	17.4	1.58
C_3H_8	61.0	102	21.7	1.67
$n\text{-}C_4H_{10}$	78.5	133	26.0	1.70
$CHCl_3$	69.9	170	30.6	2.43

V_o is the intrinsic molal volume at zero fluidity; V_t is the molal volume when the gas is expanded sufficiently to permit linear randomly oriented thermal momenta.

References: summary in ref. 6; ref. 7 and ref. 8.

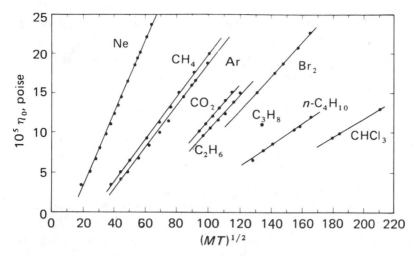

FIG. 3. Linear variations of viscosities of gases and vapors with values of $(MT)^{1/2}$. The units of MT are g °K mol^{-1}.

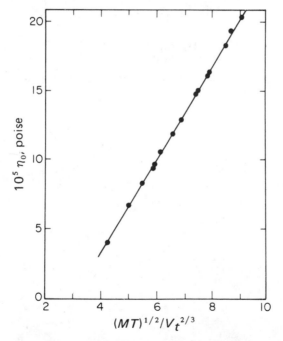

FIG. 4. Viscosities of nine dilute gases and vapors which all fall upon a single line when plotted against $(MT)^{1/2}/V_t^{2/3}$. The units on the abscissa are (g °K $mol^{-1})^{1/2}/(cm^3$ $mol^{-1})^{2/3}$.

$V_t{}^{2/3}$ that places all of their points upon the common line. This procedure will suffice to predict the whole temperature range of viscosity of a gas or vapor from a single measurement.

The equation for the line in Fig. 4 is

$$\eta_o = 3.33[(MT)^{1/2}/V_t{}^{2/3} - 3].$$

1. Hildebrand, J. H. (1969) *Proc. Natl. Acad. Sci. USA* **64**, 1331–1334.
2. Hildebrand, J. H. (1971) *Science* **174**, 490–493.
3. Hildebrand, J. H. & Lamoreaux, R. H. (1972) *Proc. Natl. Acad. Sci. USA* **69**, 3428–3431.
4. Hildebrand, J. H. & Lamoreaux, R. H. (1974) *Physica* **74**, 416–422.
5. Hildebrand, J. H. (1975) *Proc. Natl. Acad. Sci. USA* **72**, 1970–1972.
6. Person, R. D. (1958) *Kinetic Theory of Gas* (McGraw-Hill, New York).
7. Cook, G. A. (1961) *Argon, Helium and the Rare Gases* (Interscience, New York, London).
8. Landolt–Börnstein *Tabellen* (Springer-Verlag, Berlin), Vol. II 5a.

13

Correlation of the Viscosities of Liquids with Widely Different Molal Volumes and Boiling Points

J. H. HILDEBRAND, R. H. LAMOREAUX and A. G. LOOMIS
Department of Chemistry
University of California, Berkeley

ABSTRACT. Perfluoro-tributyl-amine and perfluoro-methyl-morpholine differ widely in boiling points, 177° and 50°, and in intrinsic molal volumes, 152 and 350 cm³mole⁻¹, but at temperatures where they are expanded equally over their intrinsic volumes their viscosities are equal over long ranges.

The diffusivity of light gases in the amine does not correspond with the pattern previously found for their diffusion in CCl_4, but it does for CH_4 in the f-morpholine. We attribute this to the larger spaces between very large molecules.

The fact that these liquids are moderately fluid when expanded by only 5% indicates that diffusion in liquids occurs only by thermal displacements much smaller than molecular diameters.

In 1974 Hildebrand and Lamoreaux[1] published a paper on diffusivity of gases in which diffusivities of a number of gases in CCl_4, and in similar simple liquids as well, were linear in temperature, with lines converging toward the temperature at which the self-diffusion of CCl_4 becomes zero, its freezing point. However, in the liquid perfluoro-tributyl-amine, $(C_4F_9)_3N$, the lines for diffusivity were curved and did not converge to zero at a common temperature. The present study was undertaken to learn whether this behavior is attributable to the extraordinary size and departure from symmetry of the molecules of this substance.

We inquired of a scientist in the 3M Company what perfluorochemical we might obtain whose molecules are smaller and more symmetrical, with the result that the company presented us with a generous amount of perfluoro-N-methyl-morpholine, whose structural formula is

We designate it as C_5ONF_{11}. Its properties pertinent to this research, together with those of $(C_4F_9)_3N$, are shown in Table 1. Data for the amine are those reported in refs. 1 and 2.

Table 1   Properties $(C_4F_9)_3N$ and C_5ONF_{11}

	$(C_4F_9)_3N$	C_5ONF_{11}
Density	$1.932 - 0.00216t$	$1.7785 - 0.00291t$
Boiling point	177°	50°
Volume, cm^3mole^{-1} at $\phi = 0$	350	152
Solubility parameter $(\Delta E^v/V)^{1/2}$	5.9	5.8
Vapor pressure at 25°, torr	0.55	281.7

Measured vapor pressures of C_5ONF_{11} between 5 and 30° agree within 0.1 torr with the equation, $\log_{10}P = 7.9832 - 1649.8T^{-1}$, $T = °K$.

We determined the viscosities of both liquids and report them as fluidities, ϕ , in Table 2, extending to a longer range than those reported in the former paper. Plots of ϕ against molal volumes give straight lines for all values above the dotted lines across the lower part of the table. The straight portions range from $-38°$ to $+46°$ for C_5ONF_{11} and from $31°$ to $80°$ for $(C_4F_9)_3N$. Extrapolating these points to $\phi = 0$ gives for the intrinsic molal volume of a liquid in the range where its molecules have full freedom of motion, V_0, the value of 152 cm^{3}mole^{-1}for C_5ONF_{11} and 350 cm^{3}mole^{-1} for $(C_4F_9)_3N$. The equation for these straight lines is $\phi = B(V - V_0)/V_0$. It expresses the simple concept that fluidity is proportional to the ratio of free volume to occupied volume, which we call degree of expansion.

Table 2 **Fluidities,ϕ reciprocal centipoise, of $(C_4F_9)_3N$ and C_5ONF_{11} and corresponding molal volumes, cm^{3}mole^{-1}**

	$(C_4F_9)_3N$				C_5ONF_{11}		
$t°$	ϕ	V	V/V_0-1	$t°$	ϕ	V	V/V_0-1
79.6	0.796	381.3	0.089	45.8	1.748	181.8	0.196
69.6	0.654	376.2	0.075	36.3	1.548	178.8	0.176
60.2	0.549	372.4	0.064	25.7	1.337	175.5	0.155
49.9	0.449	367.9	0.051	20.0	1.247	173.9	0.144
42.2	0.373	364.5	0.041	15.0	1.170	172.6	0.136
36.9	0.319	362.4	0.035	10.4	1.061	171.1	0.126
30.9	0.263	359.9	0.028	10.0	1.100	171.3	0127
25.8	0.222	357.7	0.022	5.0	1.028	170.0	0.118
20.0	0.179	355.3	0.015	0.7	0.951	169.0	0.112
14.5	0.144	353.0	0.009	-5.0	0.909	167.5	0.102
7.7	0.107	350.4	0.000	-10.0	0.831	166.3	0.094
	0	350.		-15.0	0.753	165.1	0.086
				-20.0	0.683	164.0	0.079
				-25.0	0.624	162.8	0.071
				-30.5	0.559	161.6	0.063
				-38.0	0.448	159.7	0.051
				-50.0	0.330	157.2	0.034
				-60.8	0.249	155.0	0.018
				-70.5	0.180	153.0	0.007
					0	152.	

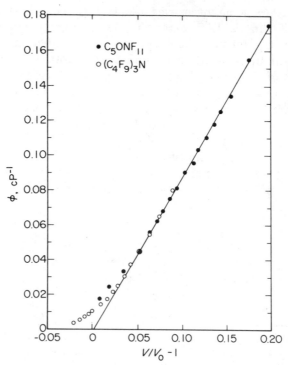

Figure 1   Relation between fluidity and degree of expansion of $(C_4F_9)_3N$ and
C_5ONF_{11}.

Values of ϕ against $V/V_0 - 1$ are plotted in Figure 1. We see the
remarkable fact that these two liquids, despite their very different
molal volumes and temperature ranges, are equally fluid (or viscous) when expanded over their intrinsic volumes. One liquid is 5%
expanded at $-38°$, the other at $+50°$. Their B-values are the same.
This accords with the small differences between the solvent powers
of perfluorochemicals. For example, the mole fractions of iodine at
$25°$ in $(C_4F_9)_3N$, c-$C_6F_{11}CF_3$, and $C_8F_{10}O$ are, respectively, 0.0023,
0.0021, and 0.0021. Iodine molecules are in nearly the same field of
potential in all these f-liquids, regardless of the size of their
molecules.

In Figure 2 the diffusivities of gases in $(C_4F_9)_3N$ are plotted as
reported in the former paper, but this time against V/V_0-1 instead
of temperature. We have determined the diffusivity of CH_4 in
C_5ONF_{11} and added the points to Figure 2. They fall on a straight
line pointing toward zero, thus behaving like gases in CCl_4. The
nonconformity of diffusivities in the amine seems attributable to

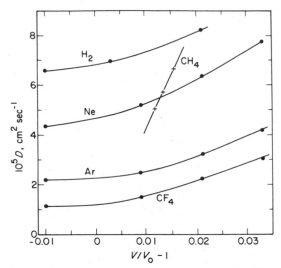

Figure 2   Diffusivities of H_2, Ne, Ar and CF_4 in $(C_4F_9)_3N$ and of CH_4 in C_5ONF_{11} at 25°, and varying degrees of expansion over V_0.

(a) measurements at temperatures extending below the straight part of the line, and (b) the larger spaces between large molecules.

It is nearly impossible to induce perfluorochemicals to freeze; isomers are usually present, and orienting forces are weak; consequently, viscosity increases to a "pour point." At the temperature of the lowest point for $(C_4F_9)_3N$ in Figure 1, $-12.3°$, it took 1233 sec. for the liquid to flow between the marks of our viscometer, at 20° it took 44.3 sec.

Finally, we call attention to the significance of moderate fluidity and diffusivity in liquids with large molecules, when expanded by only 5%. All molecules are partaking in very small but rapid thermal displacements.

REFERENCES

1.   J. H. Hildebrand and R. H. Lamoreaux *J. Chem. Phys.*, **55**, 4715 (1971).

2.   G. J. Rotariu, R. J. Hanrahan, and R. E. Fruin, *J. Am. Chem.*, **76**, 3752 (1954).

3.   J. H. Hildebrand, *Science*, **174**, 1490 (1971); also J. H. Hildebrand and R. H. Lamoreaux, *Proc. Nat. Acad. Sci.*, **69**, 3428 (1972).

4.   J. H. Hildebrand, J. M. Prausnitz, and R. L. Scott, *Regular and Related Solutions*, Van Nostrand Reinhold, New York, 1970, p. 205.